Garlic
Leeks
Celery
Asparagus
Rhubarb
Onions

PLANTS WE EAT

Garlic
Leeks
Celery
Asparagus
Rhubarb
Onions
Garlic
Leeks
Celery
Asparagus
Rhubarb
Onions
Garlic

Stinky and Stringy

Stem & Bulb Vegetables

Meredith Sayles Hughes

Lerner Publications Company/Minneapolis

Check out the author's website at www.foodmuseum.com/hughes

Website address: www.lernerbooks.com

Designers: Steven P. Foley, Sean W. Todd
Editor: Amy M. Boland
Photo Researcher: Beth Osthoff

LIBRARY OF CONGRESS CATALOGING-IN-PUBLICATION DATA

Hughes, Meredith Sayles.
Stinky and stringy: stem & bulb vegetables / by Meredith Sayles Hughes.
p. cm. — (Plants we eat)
Includes index.
Summary: Describes historical origins, use, and growing requirements of garlic, onions, shallots, and leeks (members of the allium family) as well as celery, asparagus, and rhubarb. Includes recipes.
ISBN 0–8225–2833–9 (lib. bdg. : alk. paper)
1. Stem vegetables—Juvenile literature. 2. Allium—Juvenile literature. 3. Cookery (Vegetables)—Juvenile literature. [1. Allium. 2. Vegetables. 3. Cookery—Vegetables.] I. Title. II. Title: Stinky and stringy. III. Series.
SB351.S76H84 1999
641.3'526—dc21 97–43373

Manufactured in the United States of America
1 2 3 4 5 6 – JR – 04 03 02 01 00 99

The glossary on page 85 gives definitions of words shown in **bold type** in the text.

Contents

Introduction 4
Onions 10
Garlic 28
Leeks 40
Celery 50
Asparagus 60
Rhubarb 76

Glossary *85*
Further Reading *86*
Index *87*
About the Author *88*

Introduction

Plants make all life on our planet possible. They provide the oxygen we breathe and the food we eat. Think about a burger and fries. The meat comes from cattle, which eat plants. The fries are potatoes cooked in oil from soybeans, corn, or sunflowers. The burger bun is a wheat product. Ketchup is a mixture of tomatoes, herbs, and corn syrup or the sugar from sugarcane. How about some onions or pickle relish with your burger?

First we eat, then we do everything else.

—M. F. K. Fisher

How Plants Make Food

By snatching sunlight, water, and carbon dioxide from the atmosphere and mixing them together—a complex process called **photosynthesis**—green plants create food energy. The raw food energy is called glucose, a simple form of sugar. From this storehouse of glucose, each plant produces fats, carbohydrates, and proteins—the elements that make up the bulk of the foods humans and animals eat.

Sunlight peeks through the branches of a plant-covered tree in a tropical rain forest, where all the elements exist for photosynthesis to take place.

Plants offer more than just food. They provide the raw materials for making the clothes you're wearing and the paper in books, magazines, and newspapers. Much of what's in your home comes from plants—the furniture, the wallpaper, and even the glue that holds the paper on the wall. Eons ago plants created the gas and oil we put in our cars, buses, and airplanes. Plants even give us the gum we chew.

Organisms that behave like algae—small, rootless plants that live in water

On the Move

Although we don't think of plants as beings on the move, they have always been pioneers. From their beginnings as algaelike creatures in the sea to their movement onto dry land about 400 million years ago, plants have colonized new territories. Alone on the barren rock of the earliest earth, plants slowly established an environment so rich with food, shelter, and oxygen that some forms of marine life took up residence on dry land. Helped along by birds who scattered seeds far and wide, plants later sped up their travels, moving to cover most of our planet.

Early in human history, when few people lived on the earth, gathering food was everyone's main activity. Small family groups were nomadic, venturing into areas that offered a source of water, shelter, and foods such as fruits, nuts, seeds, and small game animals. After they had eaten up the region's food sources, the family group moved on to another spot. Only when people noticed that food plants were renewable—that the berry bushes would bear fruit again and that grasses gave forth seeds year after year—did family groups begin to settle in any one area for more than a single season.

It's a Fact!

The term *photosynthesis* comes from Greek words meaning "putting together with light." This chemical process, which takes place in a plant's leaves, is part of the natural cycle that balances the earth's store of carbon dioxide and oxygen.

Native Americans were the first peoples to plant crops in North America.

Domestication of plants probably began as an accident. Seeds from a wild plant eaten at dinner were tossed onto a trash pile. Later a plant grew there, was eaten, and its seeds were tossed onto the pile. The cycle continued on its own until someone noticed the pattern and repeated it deliberately. Agriculture radically changed human life. From relatively small plots of land, more people could be fed over time, and fewer people were required to hunt and gather food. Diets shifted from a broad range of wild foods to a more limited but more consistent menu built around one main crop such as wheat, corn, cassava, rice, or potatoes. With a stable food supply, the world's population increased and communities grew larger. People had more time on their hands, so they turned to refining their skills at making tools and shelter and to developing writing, pottery, and other crafts.

Plants We Eat

This series examines the wide range of plants people around the world have chosen to eat. You will discover where plants came from, how they were first grown, how they traveled from their original homes, and where they have become important and why. Along the way, each book looks at the impact of certain plants on society and discusses the ways in which these food plants are sown, harvested, processed, and sold. You will also discover that some plants are key characters in exciting high-tech stories. And there are plenty of opportunities to test recipes and to dig into other hands-on activities.

The series Plants We Eat divides food plants into a variety of informal categories. Some plants are prized for their seeds, others for their fruits, and some for their underground roots, tubers, or **bulbs.** Many plants offer leaves or stalks for good eating. Humans convert some plants into oils and others into beverages or flavorings. In *Stinky and Stringy,* we'll explore both tasty bulbs and long, tall, stalky vegetables. The stinky part is easy—pungent garlic, savory onions, subtle shallots, and mellow leeks. Preparing a meal for dinner? These tasty bulbs are the golden ingredients that many people in the world put into the pan first, along with oil. All the stinky veggies are members of the botanical genus or grouping called *Allium* and are known as bulb vegetables. A bulb has a rounded shape, as do light bulbs or tulip bulbs, and so do allium plants. Take a look at an onion. The bulb is the underground food storage area of the plant. Most bulbs, too, feature what we call "skins," the papery outside wrappers that protect the part of the plant we eat. At the base of the bulb, you can probably still see the stem from which hang small roots. The green stalk grows above the ground but is usually chopped off as the bulbs are harvested.

Some alliums, such as onions, are biennials. In the first year, they grow bulbs. If left in the ground, they will flower and then die during the second year. Garlic and shallots, on the other hand, are perennials. Their bulbs send up a flower every spring for several years.

But what about stringy? Don't you get celery's "strings" in your teeth sometimes? Celery, asparagus, and rhubarb, unlike the bulb vegetables, are valued for what they grow above ground—their stems. Celery is good raw or cooked, either alone or mixed with other vegetables. Asparagus generally stars as a prized spring vegetable treat. And rhubarb, although a vegetable, joins with sugar to become a dessert treat for the summer or a jam for the winter.

Asparagus and rhubarb are perennials. They take a few years to get established and, once set, produce crop from the same plant for 15 to 20 years. Celery, an annual, must be planted fresh each year. This threesome, along with the bulbs of the alliums, make up *Stinky and Stringy: Stem & Bulb Vegetables.*

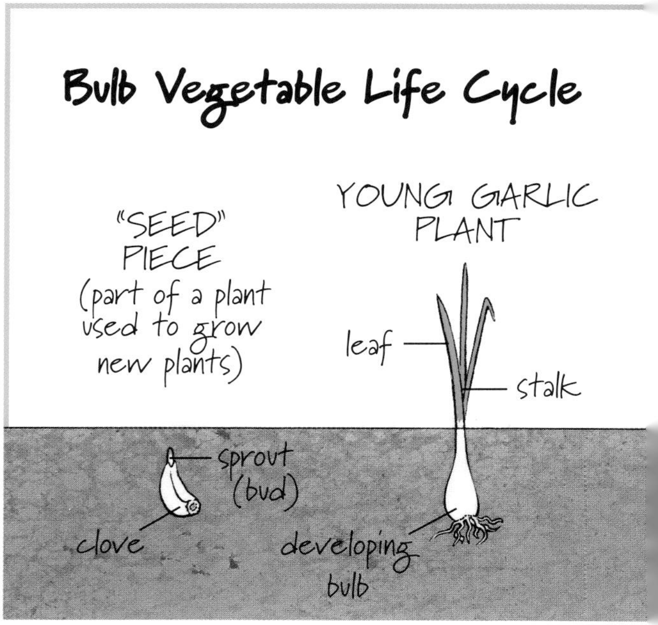

Bulb Vegetable Life Cycle

"SEED" PIECE (part of a plant used to grow new plants)
YOUNG GARLIC PLANT
leaf
stalk
sprout (bud)
clove
developing bulb

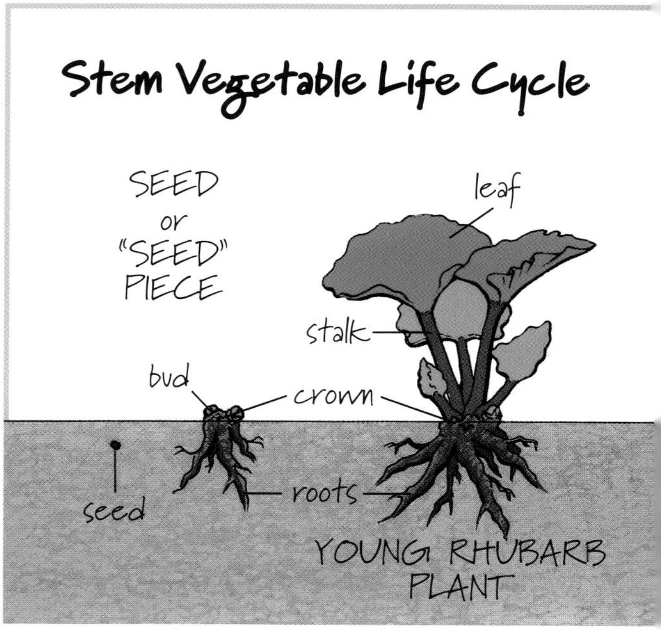

Stem Vegetable Life Cycle

SEED or "SEED" PIECE
leaf
stalk
bud
crown
seed
roots
YOUNG RHUBARB PLANT

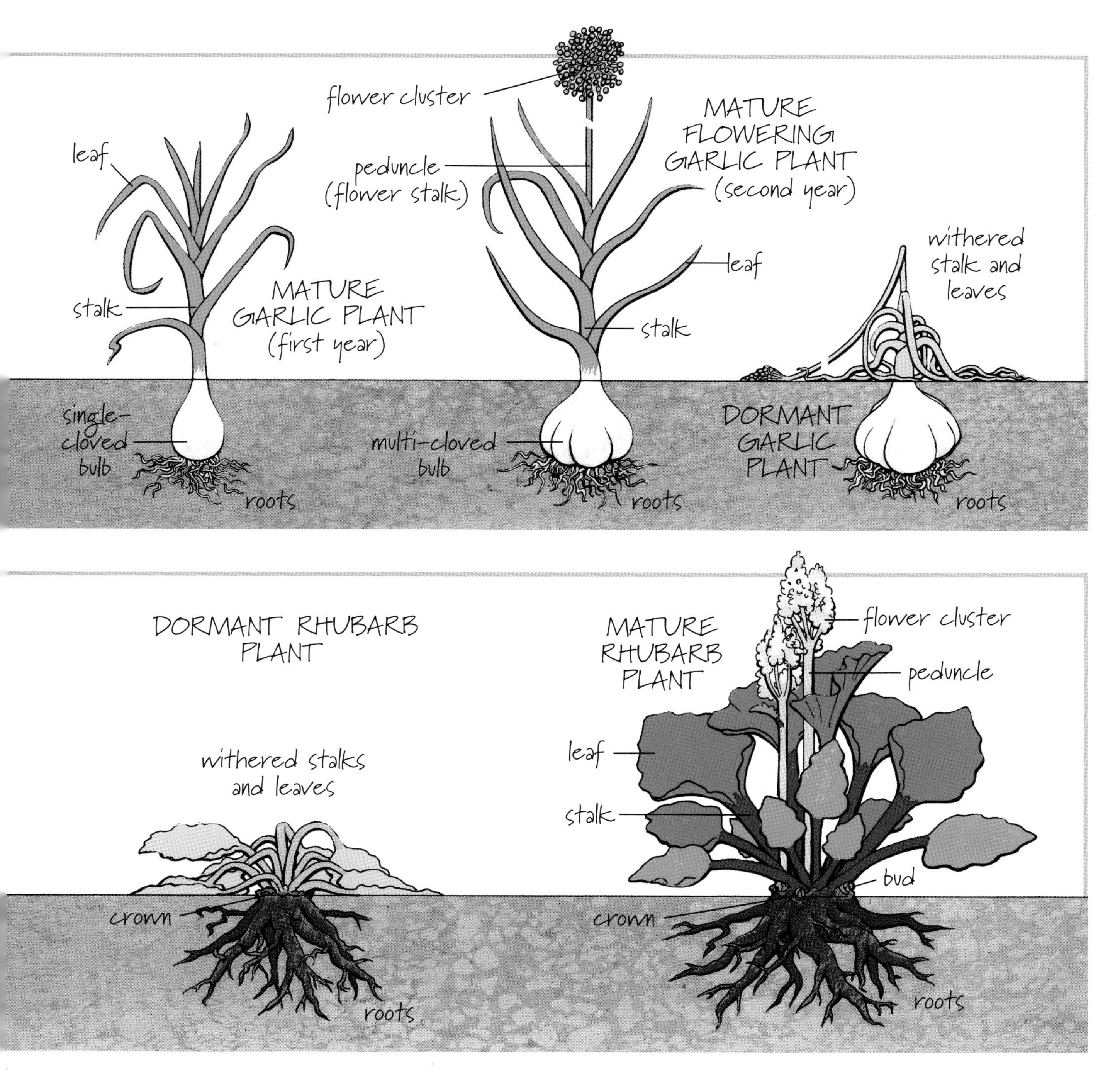
flower cluster
MATURE
FLOWERING
GARLIC PLANT
(second year)
leaf
peduncle
(flower stalk)
withered
stalk and
leaves
leaf
MATURE
GARLIC PLANT
(first year)
stalk
stalk
single-
cloved
bulb
multi-cloved
bulb
DORMANT
GARLIC
PLANT
roots
roots
roots
DORMANT RHUBARB
PLANT
MATURE
RHUBARB
PLANT
flower cluster
peduncle
withered stalks
and leaves
leaf
stalk
bud
crown
crown
roots
roots

Onions

[*Allium cepa*]

Plants of the onion family probably were being tugged out of the ground long before people knew how to grow grains and to raise animals for food. Wild bulb vegetables were easy to find by smell and, because they were fairly close to the surface, easy to gather.

Wild members of the allium family are found worldwide, and at least one prominent food historian has suggested the onion may have started life in North America. Others cite Central Asia as the source, and still others cannot decide between the Middle East and the Mediterranean region. A plant as widely spread out as the onion is an ancient plant indeed, its origins long forgotten.

Bundles of onions tempt shoppers at a street market in Honduras, a country in Central America.

Onion skins very thin;
Mild the winter
coming in.
Onion skins thick
and tuff;
Coming winter cold
and ruff.

—Old English saying

When the first onion thrust itself up out of the soil, no one chronicled its arrival because writing had not yet been invented. After writing had evolved, the onion and its relations were among the first foods to be noted. The Sumerians, who created the world's first writing, established their civilization in about 3300 B.C. in Mesopotamia, the area between the Euphrates and the Tigris Rivers, in what we call Iraq. The Sumerian legend of Gilgamesh, written down in about 3000 B.C., mentions many Sumerian foods, including a barley bread seasoned with onions. And in 2400 B.C., a Sumerian scribe (record keeper) noted disapprovingly that the city governor had the nerve to use the sacred fields around the temple for planting onions. (Onions presumably did not figure among this scribe's list of sacred foods.) Apparently kings and civic leaders as well as the poor were onion eaters in Sumeria.

Family Matters

To keep things straight in the huge families of plants and animals, scientists classify and name living things by grouping them according to shared features. These various characteristics become more noticeable in each of seven major categories. The categories are kingdom, division or phylum, class, order, family, genus, and species. Species share the most features in common, while members of a kingdom or division share far fewer traits. This system of scientific classification and naming is called taxonomy. Scientists refer to plants and animals by a two-part Latin or Greek term made up of the genus and the species name. The genus name comes first, followed by the species name. Look at the onion's taxonomic name on page 10. Can you figure out to what genus the onion belongs? And to what species?

Egyptian wall paintings dating as far back as 2700 B.C. depict onions at banquets and as offerings to the gods. The onion is the plant most frequently found in early Egyptian art, perhaps because onions were thought to represent the universe, which they imagined as made of many circular layers.

Onions turned up literally among mummies in ancient Egyptian tombs as part of the embalming process. Tomb-building workers were eating onions, too. The Pyramid of Cheops, built about 2650 B.C., bore an inscription detailing the enormous amount of onions, garlic, and radishes that fed the workers who built the tomb over a period of 20 years. Bread, onions, and beer were a standard modest meal in Egypt.

Onions were used to replace eyeballs in mummies' skulls.

Onions, as well as leeks and garlic, are specifically mentioned in the Old Testament of the Bible. During the Hebrews' 40 years of wandering in the wilderness after their expulsion from Egypt, they missed eating onions and wrote about them and other foodstuffs they had enjoyed while living in Egypt.

Starting in about 2600 B.C. sailors from Egypt, Greece, and Phoenicia (ancient Lebanon) took onions on board for lengthy sea voyages, because the vegetable was recognized early as a food that prevents scurvy. Scurvy is a disease that strikes people who don't eat enough fresh fruits or vegetables. Vitamin C, plentiful in onions, is the key to warding off the ailment.

Ancient Egyptian banquets such as this one frequently featured onions.

Perhaps onion sellers once hawked their wares on this street in Pompeii.

The ancient Chinese, whose civilization began around 2500 B.C., depended on onions, along with garlic and ginger, as basic seasonings for most of their cooking. During the same time, in neighboring India, however, eating onions, garlic, and leeks was avoided or, in some cases, forbidden. Onions in particular were thought to be bad for the eyes. But these days in many parts of India, the onion is a frequently used ingredient.

By about 1750 B.C., the Babylonians, who followed the Sumerians in Mesopotamia, produced King Hammurabi's Code, the first written body of laws. Among other things, the code decreed that the kingdom's poorest citizens would receive a portion of bread and onions each month.

Citizens of the Roman Empire—which covered much of Europe, the Middle East, and North Africa from 27 B.C. to A.D. 476—grew onions in abundance. Many people ate them daily, some say throughout each meal. Onions pickled with honey and vinegar were Roman favorites. A slice of onion on bread was a common breakfast among the poorer classes, for whom a full stomach, rather than the quality of their breath, was uppermost in mind.

Despite the fact that the onion turned up in both rich and poor households in the Roman Empire, the bulb remained downtrodden and disrespected. The Mediterranean seaside city of Pompeii, known for its fine onions, was buried entirely in ash during the volcanic eruption of Mount Vesuvius in A.D. 79. During an excavation, archaeologists found a pot of onions heavily cooked by all the hot ash. Scholars also learned that the onion sellers of Pompeii had been forced to form their own guild, or professional association, because the fruit and vegetable guild wouldn't let them in. (Maybe it was their breath!)

By the close of the Roman Empire, people living in the northern European countryside were able to get through a lean winter by consuming onions, bread, and beer. Hung high in the rafters of a cottage, onions kept well and supplied people with vitamins during months when there was little fresh food to eat. In Germany during the Middle Ages (roughly between A.D. 500 and A.D. 1500), both onions and garlic were commonly used in stews. In France wealthy landowners also appreciated onions. Some demanded a token rent of a "braid" of onions from their tenants.

Both garlic and onions are easy to braid if their long, green tops are left on.

During the Middle Ages, ordinary people like these German farmers depended on onions.

By the 1200s, onions were common in England. Through the long reign of Elizabeth I (1558–1603), onions and leeks were England's favorite veggies, often chopped and eaten with herbs as a salad. In the 1700s, sea captain and explorer James Cook insisted that his crew eat onions during long voyages. One story says that Cook refused to set sail from one port until each of his men had eaten 20 pounds of onions. Two days later, he insisted that they each consume another 10 pounds. His crews, at least, would not give in to scurvy.

Onions in the Americas

Almost all the onions grown and eaten in the United States come from the variety grown in Europe and Asia, a vegetable that settlers probably first brought to Massachusetts in 1648. At that time, however, the Americas were abundant in native varieties eaten by the locals. The Maya, who ruled Mexico's Yucatán Peninsula from the 200s until the 1500s, used a wild onion later named *Allium kunthii.* Spanish explorer Hernán Cortés noticed wild onions as he approached the Aztec city of Tenochtitlán (modern-day Mexico City) in 1519. And Chicago, the third largest U.S. city, is a permanent tribute to the aromatic bulb. Chicago is said to have gotten its name from a Potowatomi Indian word, *checagou,* meaning "place that stinks of wild onions." A different source says the Menominee Indian name *shika'ko,* meaning

Skyscrapers and busy streets cover ancient wild onion patches in Chicago.

Navajo rug weavers of the American Southwest (Texas, New Mexico, Arizona, and parts of California) have traditionally used onion skins and other plant sources for their subtle colors. Onion skins boiled in water create finely colored dyes that are perfect for yarn.

If you're not a Navajo rug weaver, you can still try this dye on hard-boiled eggs. Yellow onion skins produce an orange tint. White skins make a yellow-orange dye, and red skins will render a bluish hue.

Peel about three pounds of onions to get enough skins to fill a quart. Place the skins in a sturdy pot, add water to cover them, and bring to a boil. Boil for at least five minutes. Let cool, strain out the skins if you like, and then lower the boiled egg into the dye. The longer the egg stays in the dye, the deeper the color becomes.

"skunk place," referred not to skunks but to the odor of wild onions at the spot on the south shore of Lake Michigan we know as Chicago.

Naming Names—Onions with Star Power

People like to eat sweet onions raw and in salads (unlike the stronger "storage" onions used for cooking). So beloved are sweet onions that many varieties have risen to fame in North America. Spanish, Bermuda, Vidalia, Walla Walla, and Maui are sweet onions named for places. The Spanish onion is probably the oldest so named, but the term is used very loosely to describe both yellow and red onions. Most of the other names, however, are strictly applied to onions grown in those regions.

Famous onions didn't get their names in lights overnight. Until 1847 the Bermuda onion was just an unusually shaped, full-flavored bulb grown on the Caribbean island of the same name. But then, in one of the first trade activities between Bermuda and the United States, a shipload of these onions came to New York. They were an immediate hit in the marketplace, and soon ships made the onion passage regularly. Eventually an onion farmer in Texas acquired some of the Bermuda seeds and began to grow them at more competitive prices. These days a Bermuda onion eaten in North America is no longer likely to have come from Bermuda.

Learning from the Bermuda onion tale, growers of Walla Walla onions fiercely protect the name. Only sweet onions actually raised in the Walla Walla Valley of southeastern Washington State and northeastern Oregon can be sold commercially as Walla Wallas. This doesn't mean that home gardeners can't obtain seeds and plant Walla Wallas. Even other commercial farmers can produce the onions as long as they don't call them you-know-what.

Maui onions are another onion success story. Grown only on the Hawaiian island of that name, the Maui comes in such small quantities that you probably won't encounter one unless you visit the islands. Only 170 acres of land are planted in the onions, which sell for more than $2 a pound.

A farmer shows off his crop of Walla Walla sweet onions.

Even Vidalia's water tower pays homage to the city's sweet onion of fame.

It's a Fact!

Maui onions go through a distinct curing process that ensures their unique sweetness. Growers pull the onions and leave them in the field for two weeks. But they are not just left where they fall. The onions must all be arranged with their leaves facing the ocean. That way rainwater from nearby Haleakala, the volcanic mountain that created the island, doesn't run down and become caught in the bulb's leaves, thus causing the onions to rot.

The final sweet onion of fame in North America is the Vidalia. The state vegetable of Georgia, the Vidalia even turned up on display in Atlanta, Georgia, during the 1996 Summer Olympics. The town of Vidalia has onion stamped on it everywhere, with an onion water tower and an abundance of onion gift shops. To be a true Vidalia, an onion must be grown in one of 20 designated areas in Georgia. Farmers plant more than 14,000 acres of Vidalia each year, resulting in some 347 million pounds of onions in 1997.

One of the newest onions to make a grab for our attention is the OSO Sweet Onion. Grown in Chile, the first harvest is in early December, just in time for delivery to North American markets by the holidays. Available into March, the OSO can provide North American customers with a sweet onion in the dead of winter, a time when most other sweets are a distant summer memory.

Onions come in a variety of colors and sizes. The small, white bulbs are pearl onions. The brown ones of about the same size are shallots, which are purple inside *(inset)*.

What's a Scallion?

Some people say scallion to describe any new, young onions, also called green onions or spring onions. Others use the word to indicate the shallot. Still others think a scallion is a separate member of the onion family. Whatever they really are, scallions have straight-sided, small bulbs and tasty, green tops.

Shallots and Pearls

Even though they taste unique, shallots are really a variety of onion. What's more, the shallot's origin is obscure because it has long been confused with other onions. Shallots grow in a clump, one plant producing many bulbs, and the bulbs are usually brown skinned on the outside and pale purple on the inside. Chefs prize their subtle taste for use in special sauces and gravies, particularly the French *beurre blanc,* or white butter sauce.

Pearl onions are just tiny versions of normal-sized onions, picked early and usually packaged in small cartons and sold as a luxury vegetable. While most people seem to feel that pearls should be white, they can easily be red or yellow.

How Do Onions Grow?

We don't really know how ancient farmers grew onions, but we suspect the early steps in the growing cycle haven't changed. Onion seeds are tiny and must be raised in well-prepared, tilled areas called seedbeds where the soil is fine and rich. The seeds—scattered outdoors in the springtime or in the fall and covered with about a quarter-inch of soil—sprout quickly, within 7 to 10 days.

If the seeds fall close together, the plants must be thinned to ensure good growth of each bulb. Thinning simply means pulling out small onion plants so the remaining plants are at least four inches apart. The pulled plants can be eaten as green onions, or they can be transplanted to another prepared growing area. Commercial operations often employ vacuum planters that send seeds down tubes into several rows at once, spacing each plant correctly from the start.

After about two months, the onions are ready for harvest. The methods of harvesting differ depending on the type of onion planted and the time of the year. Sweet onions, also called fresh onions, are those generally planted in fall and winter to be harvested in the spring and early summer. Because they have a high water content, they are more fragile and require special handling at harvest. Storage onions are planted in the spring and harvested in the fall. These onions are more pungent in taste but keep well for many months, unlike the softer sweet onion.

Japanese women plant onions by hand *(above)*, spacing the seedlings so they won't require thinning before reaching maturity *(right)*.

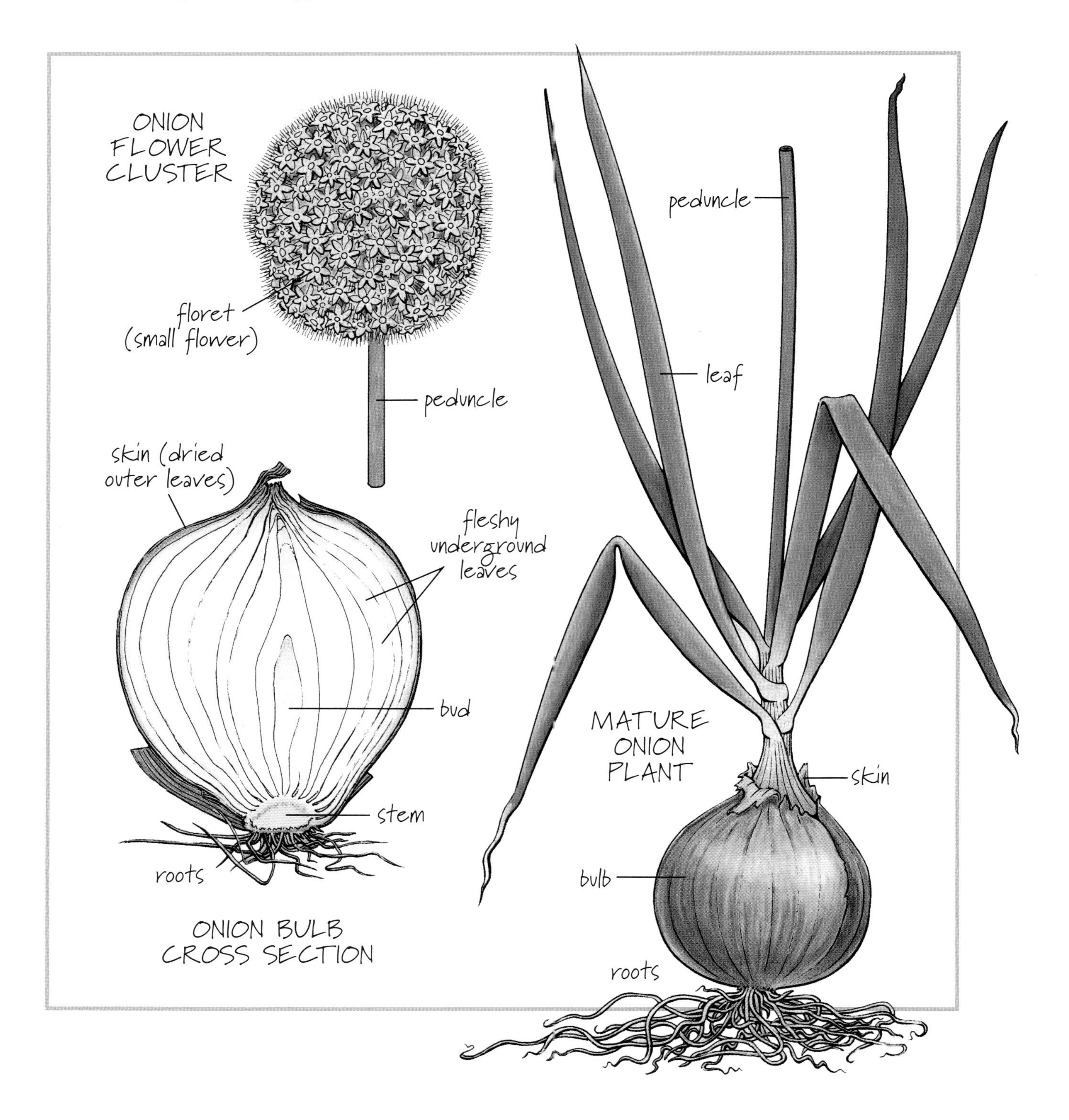
ONION FLOWER CLUSTER
floret (small flower)
peduncle
skin (dried outer leaves)
fleshy underground leaves
bud
stem
roots
ONION BULB CROSS SECTION
peduncle
leaf
MATURE ONION PLANT
skin
bulb
roots

A tractor picks up windrowed storage onions.

A picker trims an onion with her onion scissors. A large fresh-onion farm can employ as many as 250 people during a harvest.

For fresh onions, crews move through the rows, pulling up each bulb. The pickers then cut off the onion's top and trim its roots with special large shears called, appropriately, onion scissors. Workers collect the onions in plastic buckets, which they empty into burlap bags stationed in the rows. Trucks pick up the bags in two or three days.

To pick storage onions, growers use a machine called an undercutter that dives into the soil and slices beneath each bulb, separating the onion from its roots. The onions lie in the field for a couple of days until the green stalks begin to turn brown. Then a machine called a topper/loader gathers together between four and eight rows of onions in a large heap in a process called windrowing. The topper/loader picks up the onions from the pile by means of a chain belt that also cuts off their tops and further trims their roots. Workers load the onions onto trucks and take them to bins in storage buildings.

Some of the most modern topper/loaders don't require the windrowing step.

Even though its top and its bottom have been lopped off, the onion still lives. The aim of storage is to reduce the onion to a **dormant** level, much like that of a hibernating bear. In temperatures just above freezing and with humidity levels between 65 and 75 percent, onions can rest for up to six months until needed for sale.

Next the onions go to a sorter/grader, a machine that helps people to sort the onions for size and to grade them for quality. Finally the onions are bagged—sometimes in 50-pound sacks—for sale to supermarkets, which will sell the bulbs loose. Many growers also pack onions directly into 3-, 5-, or 10-pound bags for the consumer who doesn't want to hand-select each bulb.

One-third of the U.S. onion crop is sold fresh. These onions have a maximum shelf life of two to six weeks from the time they are dug. Sweet onions should always be eaten first and fast. The other two-thirds of the crop are destined for storage. Because they are harvested in the fall, these are the onions that see consumers through the long winter months.

Slice, Chop, Cook, and Eat

To describe all the ways the world eats onions would take a whole book of its own. The sizzle of sautéing onions begins meal preparations in many cultures. But onions can be main ingredients, too, not simply seasoning. For instance, onion soup (often

Weep No More!

Have you ever stood at the sink weeping while you peel and chop an onion? Your eyes water because a substance in the onion, propanethial S-oxide, mixes with the eye's natural fluids and creates a mild sulfuric acid. It's weak, but it's enough to cause people to tear up mightily. True onion lovers, of course, cannot be put off by mere tears from preparing the food they adore.

Onion people have suggested various ways to handle the bulb. One is to peel the onions under cold water in a large bowl. Another is to store onions in the fridge, then cut only the outside layers and the top part, leaving the root on until last. (Refrigerated onions lose some flavor.) Some people peel while wearing goggles. By the way—sweet onions supposedly do not cause weeping.

called French onion soup) is made from slices of onion cooked in a beef broth. In each bowl of soup floats a thick slice of bread with slightly melted cheese on top. And many U.S. families serve tiny pearl onions covered in a hot, creamy, white sauce at Thanksgiving meals. A thick slice of raw sweet onion tastes good on a burger or alone between two slices of good bread spread with butter and salt.

And of course, everyone knows onion rings, battered and deep-fried and served with a bit of salt or ketchup. A fancier version of the fried onion ring is an "onion blossom." Slicing close to the bottom but not all the way through, cooks cut the onion in eighths and gently part the sections to resemble a flower. The entire onion is battered and fried in a deep pot. People often dip the "petals" into a zesty, hot-chili mayonnaise.

Onions can also be a featured player with other ingredients in a dish. Baked squash with a rich onion and cheese filling is a staple of the Caribbean island of Martinique. Champ, or mashed potatoes with green onions and milk, is a classic Irish offering. The cheese, egg, and onion pie called quiche is a French standard eaten in many parts of the world.

Meat pies called pasties—loaded with chopped onion, potato, and turnip—come from Cornwall, England's famous peninsula. In Poland people enjoy onion croquettes.

Onion blossoms make a tasty snack.

Fresh Sweet Onion Rings with Mint
(4 servings)

1 large or two medium sweet onions
1/3 cup rice vinegar or cider vinegar
1 teaspoon sugar
1 tablespoon chopped fresh mint leaves
1/8 teaspoon crushed, dried hot red chilies
1 cup coarsely crushed ice
salt to taste

Cut the onions crosswise into 1/4-inch-thick slices. Separate into rings. In a bowl, combine the onion rings, vinegar, sugar, mint, chilies, and ice. Cover and chill for 10 to 15 minutes, stirring occasionally. Salt to taste. Serve with the ice to keep things well chilled. (But don't eat the ice cubes.)

The croquettes are made of finely chopped, fried onions mixed with cheese and wrapped in pastry flour, then coated with breadcrumbs and fried up crisp. Hollowed out, whole onions make natural baking dishes that are often stuffed with seasoned rice and cooked meat. Throughout the Middle East and the eastern Mediterranean, diners enjoy baked stuffed onions, known as *dolma* in Iraq and as *salandrouma* in Greece.

In the Central Asian country of Uzbekistan, onion bread is standard fare. Simply made from chopped cooked onions, flour, water, and butter, onion bread accompanies rice and grilled meat dishes, which are themselves often topped with raw sliced onion. Another Uzbek dish is *sabzi piez,* or braised carrots and onions. The sweet flavors of both vegetables blend together in this soothing winter dish.

Asian dishes frequently use spring onions, young plants that have just barely begun to form a bulb. Sometimes an ingredient in a stir-fry mix, finely chopped onion greens also appear as an uncooked garnish to Japan's miso soups and to Thailand's noodle specialty, *pad Thai.*

To Your Health!

Onions have many healthful applications. They contain a naturally occurring chemical, quercetin, that keeps cells from being damaged, prevents infection, and slows or stops allergic reactions. Another onion substance, adenosine, lowers blood levels of cholesterol, which can build up on the inside walls of blood vessels. Onion eating lowers blood pressure, too. And onion chewing is good for a sore throat because onion juice is a natural germ killer.

Stuffed onions are a popular dish in many countries.

Garlic

[*Allium sativum*]

Differing sources have placed garlic's origin all over Asia, and sometimes in the Middle East or in the Mediterranean. It may well be that garlic's native land was Central Asia and that travelers carried it to the Mediterranean region, the place that could be considered its secondary home. The fact that garlic has different names around the world indicates that early on it was being cultivated in far-flung areas that did not share common language roots.

We do know that more than 2,500 years ago, the Egyptians used garlic to treat headaches, tumors, and heart problems. Garlic has long been a Chinese food and medicinal staple and was written up in a Chinese medical text in 600 B.C. The Chinese probably carried the bulb to Korea and then to Japan by 30 B.C.

Each garlic bulb breaks into a dozen or more smaller pieces called cloves.

If your mother is an onion and your father is a garlic, how could your smell be sweet?

—Arabic proverb

It's a Fact!

The people of Byzantium, who in the 400s gained power over Greece and what would become Turkey, ate raw garlic crushed together with olive oil and salt.

Garlic was a favored vegetable of the ancient Greeks (1000 B.C.–400 B.C.), who preferred to eat it baked whole. The Romans connected the strength of the bulb's odor with physical strength and for that reason encouraged the empire's soldiers to eat garlic. Perhaps because of this, the Roman naturalist Pliny, who lived from A.D. 23 to A.D. 79, suggested that garlic breath could be remedied by eating an entire baked beet, if such were handy.

Garlic became standard garden fare throughout the Roman Empire, including Great Britain, where cooks regularly pounded garlic into sauces. Garlic was so mainstream that the expression arose, "the mortar always smells of garlic," presumably meaning "same old, same old." With the disintegration of the empire in 479, garlic began to lose favor in Great Britain and has not been a mainstay of British food since.

But in other parts of the waning Roman Empire, such as southern France, southern Italy, and Spain, garlic found an everlasting foothold. For several hundred years in France, a sauce called *aillee,* made from garlic, nuts, and breadcrumbs, was everyone's favorite dipping item for bread. Added to soups, aillee was supposed to be a cure for colds.

During the Middle Ages, garlic eaters proved more resistent to the dreaded plague, a disease carried by fleas, than people who did not eat garlic. The power of garlic was recognized (though not scientifically understood), and people draped their roof beams with garlands of the bulb, wore cloves of garlic in their clothing, and hung garlic on their bedposts. Waves of the plague dominated Europe in the 1300s. One bout, called the Black Death, killed a fourth of the continent's population between 1347 and 1352. Through it all, greater numbers of garlic-eating priests who tended the sick survived than did those of the clergy who refused to indulge.

Many medieval Roman Catholics felt that garlic was a scandalous food eaten by the wicked and depraved.

Roman soldiers *(facing page)*. A medieval Italian man harvests garlic *(right)*. Modern research indicates that garlic is a natural antibiotic, which may explain its effectiveness against the plague.

Go Away!!

The idea that garlic repels vampires probably stemmed from the time of the plague. Some people of southeastern Europe thought the plague was caused by evil spirits, most likely vampires. Vampires were believed to be the corpses of people not buried quite properly who rose from their graves at night to wander the countryside, sucking blood from sleeping people. People were certain that garlic made the undead creatures cringe and retreat, although we don't know why the bulb was considered so effective. (Maybe vampires didn't like garlic breath.)

Garlic's arrival in the Americas is in dispute. Some think it may have been established as a wild plant for thousands of years. A wild garlic, *Allium canadense,* grows throughout North America. Native Americans, like the ancient Greeks, ate it whole.

As early as 1774, notable American gardener (later president) Thomas Jefferson wrote in his journal that he had planted Tuscan garlic in March, along with Spanish onions. Garlic turns up regularly in garden accounts of North America from 1806 onward.

Even so, garlic was slow to make its move in mainstream North America, although it was a fixture in families originating in southern Italy, Provence (a region of southeastern France), Greece, and Spain. Not until the 1980s was garlic suddenly "in." The bulb's popularity rose in part, possibly, because U.S. students and businesspeople began traveling widely in the 1960s and 1970s. These wayfarers sampled various garlic-laden dishes, enjoyed them, and demanded garlic on their return. The same groups of people also sparked new interest in cooking international dishes at home. A rise in immigrant populations from garlic-loving Asian countries may also have played a role.

A man grills garlic bread at the huge Gilroy Garlic Festival, which attracts more than 10,000 people to Gilroy, California, each summer.

Garlic Eaters Galore!

Garlic bread may have been most North Americans' first introduction to the bulb in the 1950s, unless they came from southern European families. For years these families used garlic as a prime ingredient. By the 1980s, North Americans at large had discovered the garlic pesto sauce of Genoa, Italy.

But among the rest of the world, southern France may well be the leader of garlic cookery. After all, that region is the home of aioli, or garlic mayonnaise; *pistou,* a garlic, tomato, and cheese mix added to bean soups; and garlic herb butter made with shallots and parsley and served on grilled fish. Bouillabaisse, the fish stew from the French port of Marseille, makes liberal use of garlic. Even the simple custom of embedding slivers of garlic in a leg of lamb before roasting comes from southern France. Many of these preparations may stem from the cooking customs of the ancient Greeks and Romans who lived or traveled in this part of the world. *Skordalia,* a legendary Greek version of pesto, is a sauce made from garlic, almonds, breadcrusts, olive oil, lemon juice, and, an ingredient added in the past 500 years, mashed potato.

Spain, too, is a garlic power, using the bulb daily in many dishes. Paella, Spain's rich national dish, is made of rice and relies on garlic and saffron as primary seasonings. Another is *bacalau,* a tasty dish of baked codfish, olive oil, and garlic closely associated with the Basque people from the Pyrenees, a mountain range between France and Spain.

A Spanish waiter proudly displays his restaurant's paella.

Basque shoppers buy garlic at a market in Estella, Spain.

The French carried their garlic cookery to the islands of the Caribbean, such as Martinique. As in many of the islands, Martinique's cuisine reflects the blend of people who have made the area their home—native people, Africans, Europeans, and Asians. A dish called *bouillon à la créole* is a typical garlic-rich fish dish made with tomatoes and shallots. Other Caribbean fish sauces combine limes with garlic.

Asians use garlic widely. Deep-fried vegetable dumplings like *samosas* from India, as well as many tandoori (clay-oven–baked) dishes, rely on garlic. In the Philippines, a chopped garlic and vinegar sauce frequently appears on the table. Thai, Korean, and Chinese cooking also favor garlic. And who could overlook the often-paired Middle Eastern delights of *baba ganouche* (eggplant and garlic dip) and hummus (mashed chickpeas and garlic)?

On the other hand, garlic is never used by the people of Kashmir in northern India, who consider it somewhat dangerous—or worse, "unclean."

To Your Health!

People have long thought garlic to be good for assorted ailments. But only in recent years has science confirmed specific health claims. For example, studies from around the world have proved that garlic can reduce blood cholesterol levels. A component in garlic called diallyl disulfide (DADS) can shrink cancerous tumors. Evidently the compound changes the balance of minerals and helps to stop cancer cell growth. A National Cancer Institute study from 1992 showed that people consuming substantial amounts of onions and garlic had much lower rates of stomach cancer than did people who didn't eat the bulbs. And even without hard proof, garlic fans swear that the bulb wards off colds and swiftly halts the movement of viral infections.

Two cooks make a garlicky stir-fry, a popular Asian dish, at the Gilroy Garlic Festival. This event has helped to pull garlic out of the dried-spice cupboard and into the front ranks of foods used by U.S. cooks every day.

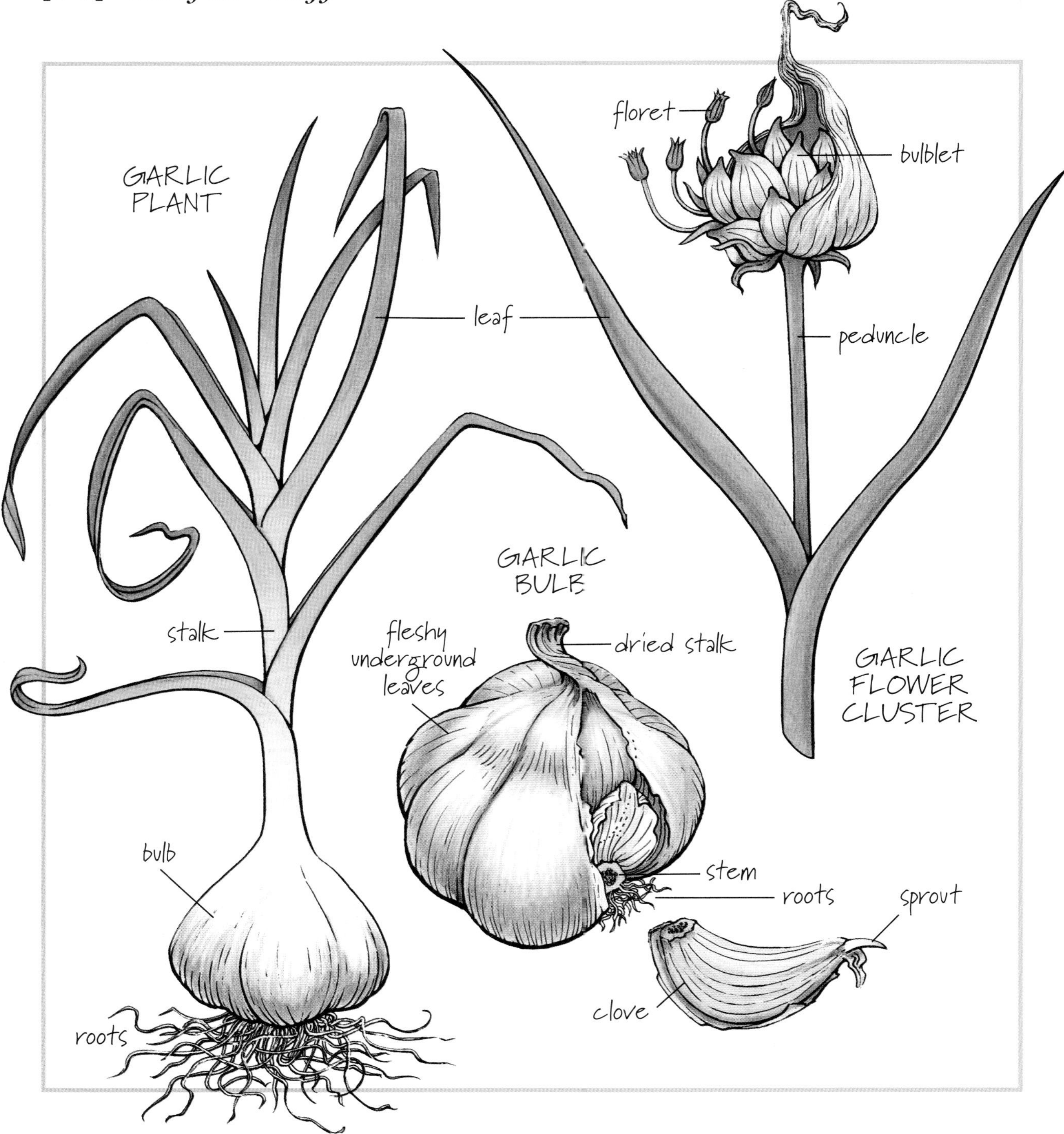
GARLIC PLANT
leaf
stalk
bulb
roots
floret
bulblet
peduncle
GARLIC FLOWER CLUSTER
GARLIC BULB
fleshy underground leaves
dried stalk
stem
roots
sprout
clove

Before long these garlic plants *(below)* will find themselves trimmed and piled into bins *(left)* at harvest time.

Growing and Selling Garlic

Ninety percent of the fresh garlic grown in the United States comes from California and is harvested mainly during May and June. While the bulb stores well, we still rely on imports to satisfy the U.S. appetite for garlic. In 1996 Americans ate more than 2 pounds of garlic per person. In 1997 garlic production in the United States was 555 million pounds, as compared to 140 million pounds in 1975. And we import more than 20 million pounds of fresh garlic as well. Argentina supplies us from January to March, when Mexican garlic takes over. Other garlic suppliers are Chile, China, Spain, and Taiwan.

Garlic is grown much like onions are, except that instead of sowing seed, farmers plant individual cloves in the fall to sprout the following spring. By the time the crop matures in the summer, no further watering is needed. After being undercut and pulled by hand, the garlic is left to cure in the field for about two weeks. Then workers trim off the roots and green stalks and bring the garlic in bins to packing sheds or processing plants.

Garlic Tech

Insects are not fond of garlic, and in the old days people used to toss cloves about the home to discourage them. You can buy a spray of 100 percent garlic juice that repels mosquitoes, aphids, whiteflies, and spidermites. Garlic cloves thrown around the base of trees and shrubs put off rabbits and deer. Dogs, though, will not be held at bay with garlic. They adore it, and garlic is a frequent ingredient in commercial dog food.

Glass jars of processed garlic, crushed or minced, are also surging in popularity among North America's busy two-income families. Even though the fresh bulb has more flavor, many people don't want to take the time to flatten and peel garlic. But garlic entrepreneurs have also created other products—like prepared salsa and pesto (an Italian sauce), garlic stir-fry sauce for Asian dishes, and garlic vinegars and mustards galore.

The Stinking Rose, a restaurant in San Francisco, California, takes its name from an old garlic nickname and features the bulb in each dish on the menu.

Garlic Mayonnaise
(2 cups)

4–6 cloves garlic
Juice of 2 lemons (a little more than 1/3 cup)
2 large eggs
½ teaspoon pepper
1½ cup olive oil

Take a broad knife, lay it on top of the cloves with the edge away from you, and lean hard on the side of the knife to crush the garlic underneath. Remove the papery covering and then chop the garlic roughly. Put the garlic and the lemon juice in a blender and run at low speed until the garlic is well minced and combined with the juice. Add the eggs and the pepper and blend at high speed. You'll need to stop a few times to scrape down the sides of the container. Turn to low speed and add the oil a few drops at a time, keeping the blender running. After a while, the mixture will become smooth and thick. Put the mayonnaise into a bowl and refrigerate until chilled, about an hour. Stir before you serve it, maybe as a topping for oven-baked French fries.

Leeks

[*Allium ampeloprasum* var. *porum*]

The leek is the statuesque member of the allium family, with tall, spreading, flat, green leaves springing from a tall, white bulb. This bulb, unlike those of onions or garlic, is straight and even rather than round. Valued for its mild but distinctive taste, the whole leek, in contrast to most of its fellow alliums, is edible, leaves and all.

While the leek is often described as native to the Mediterranean, some historians feel its tolerance of cold and its lack of resistance to heat indicate an origin farther north. The first written reference to the leek is in a 4,000-year-old record from the ancient Mesopotamian city of Ur. But it's possible the plant did grow in colder climes, and an unknown people may have brought leeks to the Middle East well before recorded history.

Harvested leeks lie beside growing plants on a British farm.

Well loved he garlic, onions, and also leeks . . .

—Geoffrey Chaucer

In any event, leeks were a favorite vegetable of the ancient Egyptians. The vegetable later found its way into the hands of the Greeks and Romans, who felt that leeks were good for the throat. The Greek philosopher Aristotle, who lived in the 300s B.C., claimed that partridges' loud, clear calls were directly linked to their love of leeks. Another early celebrity known to be a "leek head" was the Roman emperor Nero who, in the first century A.D., ate vast amounts of leek soup to improve his speechmaking voice. His nickname, Porrophagus, means "leekeater."

Even into the twentieth century, people used leek syrup as cough medicine.

Paddy's Leeks

There's an old Irish tale of a sick, elderly lady who related to Saint Patrick, the patron saint of Ireland, her vision of a special herb she must eat or die. The goodly saint asked her what the herb was. "It looks like rushes," she said. (Rushes are tall, reedy grasses that grow along the banks of streams.) Saint Paddy seized some rushes, he turned them into leeks, and the woman ate. Lo and behold, she immediately regained her good health.

Nero ruled Rome from A.D. 54 to A.D. 68.

Wales, on the island of Great Britain, has a cool, moist climate and may have been the leek's original home.

About 2,000 years ago, when the ancient people called Celts first began their journeys westward from northern Europe, they may have brought the leek with them and planted it throughout their domains, particularly in the British Isles. The leek has long been associated with the Welsh and the Scots, both descendants of the Celts. Wild leeks still grow along the southern coast of both England and Wales. In fact the Welsh consider the leek their national symbol. In Wales there is a traditional spring plowing festival called the *cymmortha,* a gathering as old as the Welsh people themselves. After the plowing, it was customary for each farmer to contribute a homegrown leek to a communal pot of stew.

It's a Fact!

The Welsh word for daffodil, *cenhinen pedr,* and the word for leek, *cenhinen,* are exceedingly similar. Some historians have decided that the leek as symbol is an English translator's mistake. In any event, both plants are bulbs, each grows wild on the islands off the southern Welsh coast, and both share the spotlight as symbols of Wales.

So it was only natural that in A.D. 640, when the Welsh king Cadwaladr defeated his enemies in battle, his troops were wearing leeks stuffed into their caps. This was the era before fancy uniforms, and the Welsh wanted to distinguish friend from foe in the heat of combat. Legend claims that none other than David, the patron saint of Wales, advised soldiers to wear the leek. Since then it has been customary to wear a leek on St. David's Day, March 1. Welsh armies continued to adorn themselves with leeks at least into the Middle Ages, and soldiers serving in Welsh military regiments supposedly still eat raw leeks on St. David's Day.

Welsh soldiers frequently carried leeks into battle.

My Soup Sprang a Leek

In the British Isles, ancient home of the Celts, leeks turn up in cockaleekie soup, a distinctive dish of Scotland that certainly sounds like a combination of rooster and leeks but is actually made with hen. Elsewhere in the world, the leek is probably best appreciated in France, where it has found favor since the Middle Ages. Leeks are a fixture of France's beloved *potage Parmentier,* a rich, creamy, potato-leek soup.

Cockaleekie soup

Baked leek in cream

Elephant Garlic?

Elephant garlic, which looks like a fist-sized garlic bulb, is not really garlic. (Nor is it an elephant, so it's safe to allow into your house with the other groceries.) Elephant garlic is actually a leek in garlic's clothing. The bulb is vying for popularity as a less pungent alternative to garlic and is sometimes even served on its own as a vegetable accompaniment to a meal. You might be the perfect elephant garlic customer if you want just a hint of garlic when you cook. Keep in mind, however, that you're eating a leek.

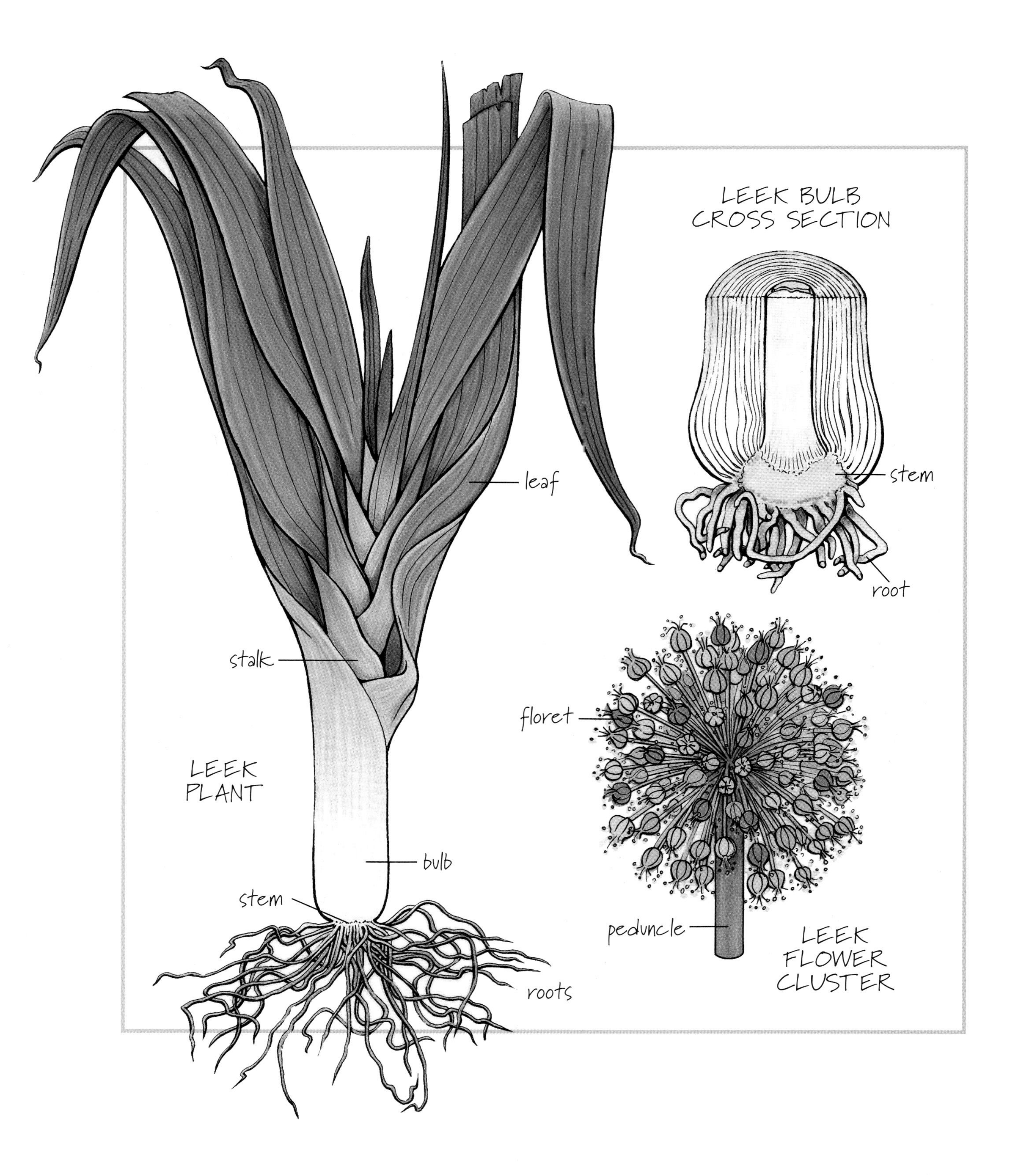
LEEK BULB
CROSS SECTION
stem
root
leaf
stalk
LEEK
PLANT
bulb
stem
roots
floret
peduncle
LEEK
FLOWER
CLUSTER

Growing Leeks

Leeks are usually planted in the fall for summer harvest, just like fresh onions. But growers can also start leeks as seed two months before the last frost of the winter. The resulting five-inch baby plants can go into the ground in early spring for an October harvest. Leeks are also harvested exactly like onions except for one detail. When the leek is about the size of a pencil, the soil is often drawn up around it to ensure a long, white portion. This step, called **blanching,** keeps the leek pale in color because with no sunlight reaching the plant, photosynthesis and greening cannot happen.

Leeks can stay in the soil all winter and still be good to eat before the ground begins to warm. As a matter of fact, these vegetables store best if left buried.

In North America, small amounts of commercially grown leeks are available in supermarkets. For the most part, though, the leek is the province of the home gardener.

Trays of leek seedlings *(top)* await planting by crews of farmworkers riding behind tractors *(bottom)*.

Eek, It's a Giant Leek!

If you'd like to grow a veggie the size of a fire hydrant, then join the ranks of giant leek gardeners. These huge plants can reach more than 3 feet in height and can weigh 12 pounds.

Each potential giant requires a 3-by-3-foot growing space with soil at least 12 inches deep that is full of **organic matter.** Most gardeners buy their seeds from specialists who feature giant plants. First sprout the seed in a pot indoors and nurture it to an ordinary 5-inch height. Then transplant the leek into the outdoor plot. Surround the plant with a circular fence of about 14 stakes or bamboo canes. Shove each cane into the soil about 1½ feet from the seedling, keeping the canes about 6 inches apart. Connect the canes to one another with a piece of string tied a little bit higher on each. As the leek grows, this framework will support the heavy leaves.

Feed the leek each week with a nitrogen-rich liquid fertilizer and give the plant plenty of water. When the giant is ready for harvest, take great care in removing it from the soil, as its root structure counts toward its total weight. Dig it out of the ground as if it were an ancient Egyptian artifact. When dug, display it! Then astound your neighbors as you feed them all baked leek in cream with grated cheese, made from one leek!

Potage Parmentier
(4–5 servings)

3 medium-sized leeks
3 medium-sized potatoes
4½ cups chicken broth
1¾ cups cold water
½ cup half-and-half or whole milk
2 tablespoons butter
2 teaspoons salt
¼ teaspoon pepper
chopped chives to garnish

Wash the leeks thoroughly and slice them one-eighth inch thick. (Do not use the tough, dark green part.) Peel the potatoes and slice them one-eighth inch thick. Combine the leeks, potatoes, broth, and water in a large, heavy pot or saucepan and cover. Bring the mixture to a boil over medium-high heat. Reduce heat and simmer 35 to 45 minutes or until the vegetables are tender. Without draining off the broth, mash the vegetables in the saucepan with a potato masher until they are fairly smooth. (If they will not mash easily, the soup has not cooked long enough. Let it simmer 10 to 15 minutes longer.) Add the half-and-half, butter, salt, and pepper. (You can add up to ½ cup of skim milk if you want thinner soup.) Heat the soup just to the boiling point. Be careful not to boil it, though, because boiling will make the half-and-half curdle and form lumps. Sprinkle each serving with chives—another member of the allium family whose green tops are used as an herb.

Celery

[Apium graveolens]

Celery was once associated with the gloom of the tomb—the Romans liked to use it to decorate coffins at funerals. But they also felt that wearing crowns of celery helped to ward off headaches after hard nights of partying.

Celery is descended from an ancient wild plant that still grows in salty marsh areas and enjoys cool, moist, black soil known as muck. Probably originally from the Mediterranean, wild celery is also found in the British Isles, throughout Europe, from western to eastern Asia, in northern and southern Africa, and in South America. Celery's earliest travels beyond the Mediterranean are not documented. Food historians assume that well-traveled Roman officials likely carried the plant throughout the ancient empire.

Since ancient times, celery has been a popular vegetable.

Celery, raw,
Develops the jaw,
But celery, stewed,
Is more quietly chewed.

—Ogden Nash

Celery was often a welcome guest at Roman parties.

Celery leaves were a favorite image of French stone carvers working on Gothic cathedrals between 1200 and 1400.

One food historian believes that even though the Romans grew domesticated celery, by the time of the Middle Ages, people had forgotten how. Wild celery must have been abundant, so many people probably picked the strongly flavored wild plant.

People originally used wild celery as a flavoring in soups and stews rather than as a vegetable to be enjoyed on its own. The French and the Italians eventually bred wild celery into the plant we know these days. And yet no written mention of celery appears in France until 1641, when celery was recorded as grown in King Louis XIV's garden. In fact the king's gardener may have been the first to have the notion of blanching celery. This practice continued up to the end of the 1800s in much of Europe and North America and is still done in some areas.

It's a Fact!

Smallage, meaning "small parsley," was the ancient English name for wild celery. (Parsley is related to celery.) The word *celery* is derived from the French *seleri,* which came from the Greek *selinon,* a plant that the Greek writer Homer mentioned in *The Odyssey* in 850 B.C.

Celery Plus

A subspecies of celery called celeriac *(Appium graveolens rapaceum)* is grown for the swollen base of its stem rather than for its stalks. The large stem, which resembles a rough-coated turnip, is often grated for use in salads or peeled and boiled for an addition to soups or stews. Celeriac tastes just like celery. Popular in France and Belgium, the vegetable is beginning to catch on in North America, too.

In China and India, farmers grow a wild celery variety for its tiny seeds, 750,000 of which make a pound. Used to flavor salad dressings, sauces, soups, and vegetable juices, celery seeds are also finely ground and added to spice mixtures. In North America, celery seeds grown commercially for raising new crops sometimes lack the proper amount of certain parts necessary for planting. Growers sell these seeds to the spice trade.

A farmer packs her celeriac, a vegetable related to celery, for market.

Growing Good String

Celery can be a difficult plant to grow. Because it has a shallow root system, celery needs a great deal of steady moisture. The plant takes loads of time to mature—as much as five to six months—during which there must be no frost. Growers have a short window of opportunity—only six to eight days—to harvest celery and bring it to market at its peak. Too much rain can flood fields and slow the harvest, and the delay will mean poor-quality celery that cannot be sold.

California is the U.S. celery giant, producing some 1.65 billion pounds—that's 83 percent of the U.S. total! Florida trails behind at 200 million pounds, and Michigan is a distant third at 103 million pounds. In the rest of the Americas, Canada, Mexico, and Guatemala also raise celery. China is a substantial celery grower, as well as the largest importer of California celery. The Netherlands, Belgium, and Britain are Europe's main celery growers.

Growers start celery seeds indoors in flats (shallow, dirt-filled boxes) or other special units, as the tiny seeds are tricky to plant directly in the field. About 10 to 12 weeks after sprouting, workers transplant the three- or four-inch-tall seedlings outside. They place the plants about seven inches apart and add lots of fertilizer and water. To ensure a successful crop, celery requires more attention to water and feeding than do most other plants.

In recent years, because of bad weather, Michigan, which has only 8 percent of the celery fields in the country, has received 73 percent of the disaster money available to celery growers.

The Tale of Michigan Celery

From the 1850s until after World War II (1939–1945), Michigan was a major celery grower. Kalamazoo, Michigan, was known as the Celery City, and the distinctive odor of celery could be smelled all around town.

The first celery raised in Kalamazoo was not the type we eat these days but rather a hollow-stalked variety best used in soups. In 1866 a seed grower named Cornelius de Bruin spotted a somewhat different-looking plant among the soup celery in a local garden. He asked for seed from the plant and grew some on trial. Soon he and his brothers were in the eating-celery business.

By the 1880s, Celery City was selling not only fresh stalks but also celery-based health remedies, celery soup, celery mustard, and celery ade—a forerunner to Dr. Celray's Soda, a celery-flavored soft drink still found in delis.

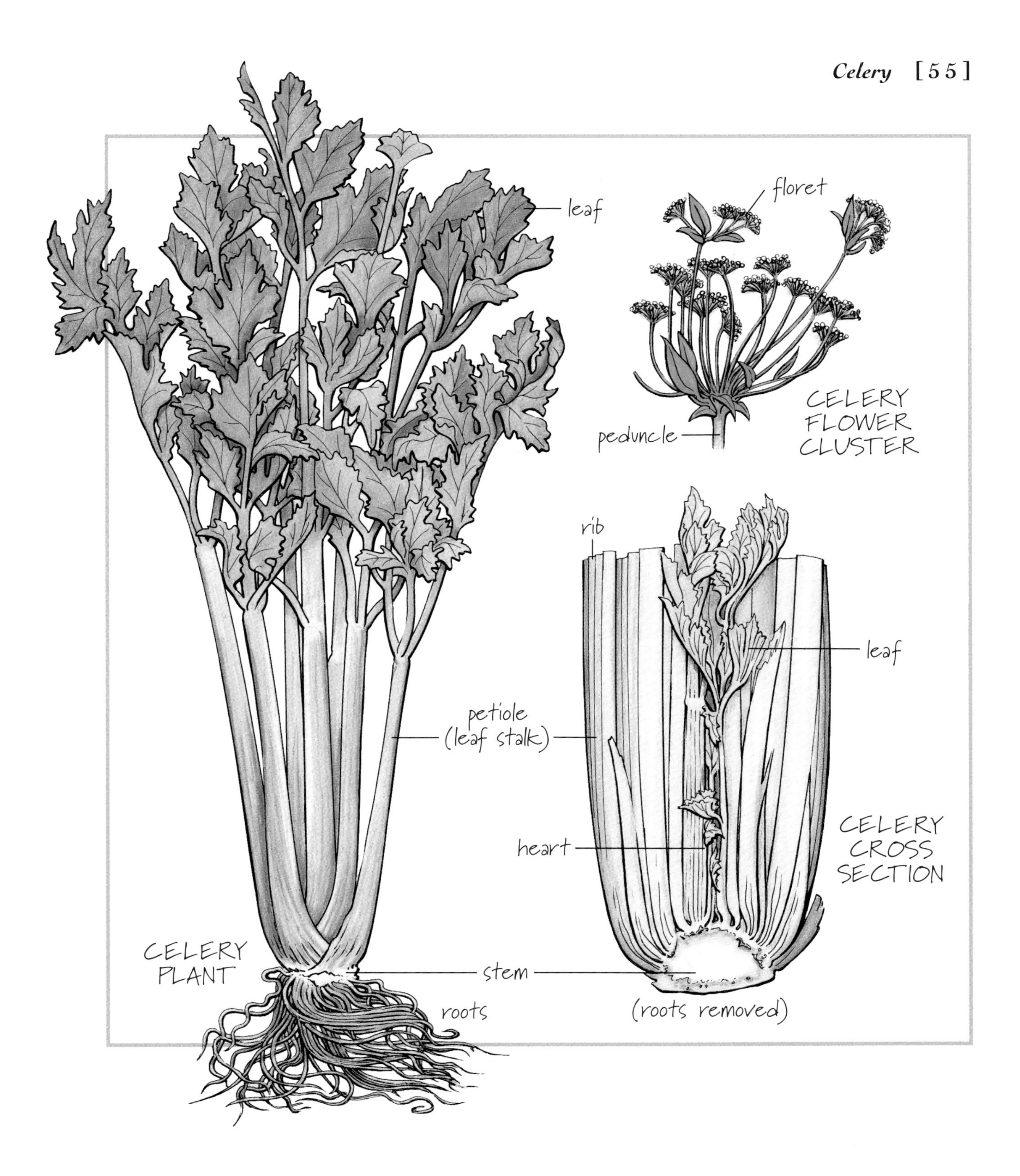
leaf
floret
CELERY FLOWER CLUSTER
peduncle
rib
leaf
petiole (leaf stalk)
heart
CELERY CROSS SECTION
CELERY PLANT
stem
roots
(roots removed)

While the celery is growing, a few farmers may blanch it to keep it pale. Workers pile soil or place paper cones around the stalks to stop sunlight from reaching that part of the plant. Newer varieties of "self-blanching" celery do not require either method, as their stems will not turn green even in sunlight. For the most part, though, modern consumers feel that naturally green celery is more nutritious, so the market for the blanched product in North America is extremely limited.

After five or six months, when the stalks are at least six inches tall, celery can be harvested. The harvesting machines have two belts that straddle the plants along each row. The belts lift each plant, slicing off the top and bottom as the machine moves along. The celery is then taken by truck to the processing area where the hand labor begins. Workers pick the celery off a conveyor belt, strip the outside stalks off each bunch, and even up the top and bottom cuts made by the machine. A machine called a washer removes all the field dirt. Then the bunches are packed by size in cartons for shipment either directly to a supermarket chain or to a food broker, who will sell the celery for the farmer.

To blanch celery, some farmers place paper tubes around the growing plant.

A celery-harvesting machine mows its way through a field of mature crop.

A Guatemalan farmer bundles celery to sell at a local market.

Capillary Action

A celery stalk is a perfect tool for demonstrating capillary action—the movement of liquid within a solid. The word *capillary* comes from the Latin word for hair. Chop off the bottom of a celery stalk and examine it closely. You will see the tiny hairlike tubes that the stalk uses to draw water and minerals up from the ground through the stem and into the plant. Fill a short clear glass with lukewarm water and add several drops of food coloring. Insert the prepared celery stalk and wait 30 to 40 minutes. When you look again, you should see color all the way up and down the tiny tubes. (The color is easier to see if you peel away the celery's skin or thin first layer.)

In the 1890s, soothing drinks called tonics were big sellers. Most of these started as home remedies, herbal concoctions made up and used within families. Paine's Celery Compound from Vermont was one such tonic, along with Kalamazoo Celery Nerve and Blood Tonic.

Folklore has long held that celery lowers blood pressure, and research is proving the claim. William Elliott, a scientist at St. Luke's Medical Center in Chicago, Illinois, discovered a compound in celery known as phthalide, which increases blood flow by relaxing the arteries. Elliott's research showed that people may be able to lower their blood pressure noticeably by eating four stalks of celery. When blood pressure is normal, people feel relaxed instead of tense and hyper and also have a lower risk of heart disease.

Cooking and Eating

One British recipe for celery "ice cream" combines puréed celery, cottage cheese, and sour cream to be served in the summertime with cold chicken or fish! Most people would rather eat celery in more familiar ways. North Americans may associate celery with the Thanksgiving Day meal—pale green stalks crisply cool in ice water or yellowish leaves chopped up into the turkey stuffing. Or perhaps creamy celery soup. And don't forget Waldorf salad. Invented sometime during the 1890s at New York's famous Waldorf-Astoria Hotel, this salad features celery, apples, walnuts, and mayonnaise served on a bed of lettuce.

There's a celery joke going around that it takes more calories to eat a stalk of celery than are contained in the stalk itself. Maybe it's true—two stalks have only 25 calories.

In French cooking, chopped celery, onion, and sometimes carrot are often the start of a chicken stew or a vegetable soup. The Chinese use celery in stir-fries but also as part of a salad—chopped fresh celery with dried shrimp. A relish from Georgia, a country on the Black Sea that was once part of the Soviet Union, blends grilled mashed eggplant with chopped celery leaves. The Flemish dish known as *waterzooie,* associated with the Dutch-speaking people of northern Belgium, is a thick soup made with either chicken or fish and chopped celery. Another Belgian favorite is braised celery, which the French love, too. Cooked slowly in butter, lemon juice, and a bit of water, the celery stews in its own juices, becoming soft and acquiring a rich, sweet flavor.

Celery baked with ham and Parmesan cheese is a rich Italian dish. Another somewhat lighter food combines chopped celery with potatoes cooked in olive oil and lemon juice. Greeks like to roast celery with pork, topping it all with a sauce made of egg, lemon, and pork drippings.

Celery Is Child's Play

Consider "ants on a log," a favorite recipe of those too young to be allowed to turn on the stove. Take one celery stalk. Fill the hollow with low-fat cream cheese or peanut butter. Place "ants," otherwise known as raisins, along the top. Observe the ants silently. Chomp down.

Asparagus

[*Asparagus officinalis*]

Asparagus is a bizarre-looking plant. Its long, slender, new shoots (the part that we eat) resemble spears. If left unharvested, the spears develop into ferny leaves. The plant grows red berries that carry the asparagus seed. Birds love to eat the berries, and they deposit seeds in their droppings far away from the original plant. Because of this, asparagus pops up not only alongside areas where it has been cultivated but also much farther afield. Food experts assume that birds long ago carried the plant's seeds from their original home to parts of the Mediterranean and North Africa. But the experts can only guess as to the birthplace of this ancient plant.

Freshly picked asparagus spears lie in a bin, waiting to be taken from the field.

The green and gold of my delight—Asparagus with Hollandaise.

—Thomas Augustine Daly

Some scholars say the plant first grew in coastal areas around the Mediterranean, because wild asparagus plants still thrive there. Others propose the British Isles, or perhaps the coastal area along the Baltic Sea in northern Europe. The northern origin seems more likely because asparagus likes to shut down fully during the growing season, usually during a winter freeze. Whatever its beginnings, since ancient times asparagus has been prized as an extravagant treat.

Soldiers and traders of ancient Rome may have carried asparagus throughout the empire—if the birds had not already done so. Wild asparagus still grows in Provence and on the Mediterranean islands of Corsica and Sardinia, all once part of the Roman Empire.

Romans wrote about the vegetable, but after those accounts, there's little mention of asparagus cultivation until about 1300. Then records indicate that farmers near Paris were

Asparagus has to be cooked swiftly until tender but still crunchy, as even the Roman emperor Augustus, who died in A.D. 14, knew. He coined a Latin phrase—*velocius quam asparagi coquantur*—that translates as "faster than you can cook asparagus," meaning "in a flash."

growing asparagus. Food historians assume that even if domestic cultivation had died out for a while, people simply continued to eat the wild version of the plant.

Asparagus was well established throughout western Europe by the Renaissance (1300–1600). British settlers carried the veggie to North America in the 1600s. Instead of seed, they brought **crowns**—the plant's first-year growth of roots and undergound stem—from which asparagus is usually grown.

The Greeks probably coined the word asparagus, which means "young shoots."

Asparagus crowns provide a crop more quickly than does growing the vegetable from seed.

Asparagus sprouts *(inset)*, if allowed to grow, become tall, fernlike plants.

Growing Spears

Growing asparagus is not for the impatient. The first year, one must stand back, watch the ferns develop, and avoid picking any of the spears, no matter how tempting. This is because the asparagus plant, while a long-lived perennial, needs a good start to establish a permanent root system. As part of its growing cycle, the plant also requires a period of dormancy (rest), which usually occurs in winter in northern climates. After three to four years, the well-tended asparagus plant can produce a good crop for as long as 15 years. So it is a plant for long-term planners.

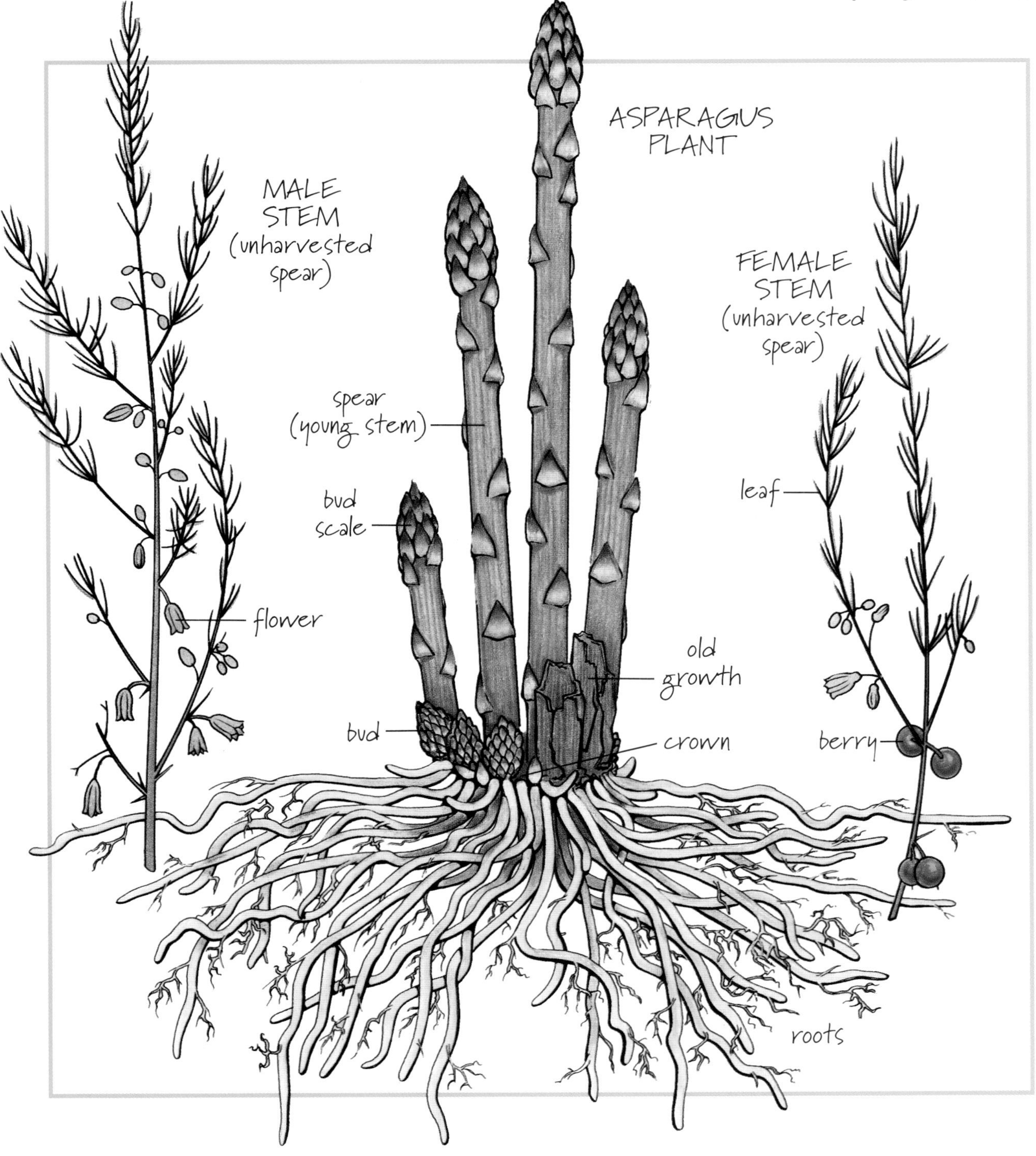
ASPARAGUS PLANT
MALE STEM (unharvested spear)
FEMALE STEM (unharvested spear)
spear (young stem)
bud scale
flower
old growth
bud
crown
leaf
berry
roots

No More Co-Ed Growing

Until the last 10 years or so, it was customary to grow both male and female asparagus plants in the same field, since both produce the spears we eat. In late spring, male and female plants produce flowers. The pollen from the male flower fertilizes the female flower, which develops berries. When the berries go from green in summer to red in the fall, farmers collect the fruits, extract the seeds, and plant the seeds in nurseries to produce the crowns most people plant when starting an asparagus crop.

But female plants cause trouble for commercial growers. Any uncollected berries drop to the ground and the seeds within them sprout, sending up seedlings. Because these weedy plants, called "volunteers," grow densely and close to the mother plant, they can attract unwanted insects and a range of plant diseases. So many growers use all male plants.

While the plant can be grown from seed, this adds yet another year to the process. Instead most people buy crowns from reputable suppliers. Growers place the crowns about 12 to 16 inches apart in well-drained soils in 6-inch-deep furrows (V-shaped trenches), then cover the plants with about 3 inches of soil. As the plant grows, more soil goes into the trench so that by the end of the first growing season, the furrow is filled.

Finally by the spring of the second season, a few of the spears are available for cutting or snapping off. Harvesters cut off the spears with a knife about one inch below the soil line, taking care not to damage the crown. Spears snapped off by hand will not be as long as cut spears. An average spear is about 7 inches in length, but the asparagus plant's spears can shoot up as much as 10 inches in a 24-hour period!

Asparagus is harvested by hand whether grown in the home garden or commercially. Although there have been monumental efforts to develop effective machine harvesters, no machine has yet filled the bill as well as a human picker. That's because asparagus spears shoot up at different times and achieve various heights. In commercial operations, three to seven people sit on a low platform that is slowly pulled through the rows by a tractor. As they ride through the rows, the farmworkers select spears of similar height and pick only those. Later they can make a second round and cut the shorter spears. This kind of selection is much harder for a machine to pull off. In North America, crews using this method enter the fields about 20 times to bring in the season's single harvest.

Pickers ride behind a tractor.

Spears intended for the fresh market are sorted by size, packed in crates with the shoots' tips pointing up, and then cooled to aid in storage before shipping. Asparagus for canning or freezing travels in bushel boxes to processing plants for weighing and cooling in cold water tanks. Within 24 hours, the spears are ready to go to supermarkets.

Packers sort asparagus for shipment.

In the spring, German restaurants often offer their customers two menus. One shows standard fare, and the other lists a range of asparagus dishes, sometimes as many as 20 in all.

Global Growing and Selling

In the Western Hemisphere, the United States and Peru—located on the Pacific coast of northern South America—both put out about 200 million pounds of asparagus a year. China is the world's biggest asparagus grower, producing about 5.5 billion pounds a year and supplying the bulk of canned asparagus sold in Europe. In Africa, South Africa is the biggest producer, with Morocco also important as a supplier to European markets. Spain (with Greece in second place) is the top-ranking grower in Europe, and Germany is probably Spain's biggest customer. Germany and France are huge consumers of asparagus

White Asparagus, Anyone?

Although things are changing, European asparagus consumers have traditionally preferred white (blanched) asparagus. Growers bury the crowns of these asparagus with as much as 13 inches of soil so that the spears are completely covered as they grow. As soon as the tips of the spears are visible, the asparagus is picked by hand by poking a knife into the mound of dirt. Because the spears receive no light, photosynthesis, which causes the stems to turn green, cannot take place. Originally green asparagus was seen as bitter, and the white as sweeter and tastier. But plant breeders have eliminated any lingering bitterness from green varieties. Even though it's probably not true these days, Europeans still consider the white asparagus to be a bit more tender and somewhat more moist. And traditional eating habits have staying power.

products of all kinds, both green and white, processed and fresh. Japan is most likely the world's largest importer of asparagus, buying from the United States, Australia, the Philippines, Thailand, and New Zealand.

While asparagus traditionally has been grown in northern areas of the world, the vegetable's "luxury" status has sparked efforts to grow it in other regions. The veggie's prestige stems from the fact that historically it has been a food available only briefly each year and thus has been anticipated eagerly. Asparagus eaters have been willing to pay more for their vegetable because of this "right now or next year" situation.

Peru is one of the world's new asparagus powers. Peruvian farmers are irrigating high-altitude desert lands to grow the crop. Because the area does not have a winter freeze to bring on the plant's dormant period, growers simply stop watering the plants and let the top fernlike greenery die off. After mowing away the dried-up ferns, the plants are re-irrigated until new spears push through the ground. Because the temperature is constant year-round, using this method produces two or more harvests per year.

Asparagus also must be grown without winter dormancy in southern Taiwan, a humid area of the **tropics** off the southeastern coast of China. Taiwanese farmers have found a way to make the asparagus produce shoots throughout the year. When the plant sends up spears, two are left to grow into ferns. The rest of the spears are harvested as they come up. As the ferns begin to die back, two more spears are allowed to fern out, taking their place. In this manner, the plant is able to continue producing over a relatively long period of time.

A Taiwanese farmer shows off his crop.

Although there are dozens of ways to cook this vegetable, many people still consider plain spears topped with Hollandaise sauce to be the classic asparagus dish.

Spearing a Few

Fans of fresh food tend to eat asparagus simply boiled and utterly plain, but the history of asparagus cooking is rich. From the beginning, this vegetable has been sauced or wrapped or otherwise hooked up with other foods.

Until the late 1800s, many English-speaking people called asparagus by the nickname "sparrow grass."

In western Europe, the United States, and many other places, you will often find asparagus doused with Hollandaise sauce, a creamy concoction made from lemon juice, egg yolks, and butter. Another delicious preparation appreciated in these places is asparagus sprinkled with chopped hard-boiled egg and melted butter. Cooks also

To Your Health!

Asparagus has a powerful cancer-fighting compound called glutathione, which seeks out and neutralizes poisonous chemicals.

From its beginnings as a crop, asparagus has been considered a healthy food and "good for the blood." While all vegetables give us some nutrients, asparagus provides an amazingly well-balanced range of substances. Included are potassium, which keeps our water levels in balance; fiber, which keeps our digestive systems running smoothly; vitamin B_6, which helps make the best use of our food; and thiamin, important for daily energy. Of all vegetables, asparagus has the highest content of folic acid, the key component necessary for blood cell formation. Asparagus has no fat and very few calories, about four per spear. If you are selecting a green vegetable to eat for maximum food value, go with asparagus.

Asparagus is a versatile and delicious vegetable, whether served alone *(left)* or cooked with other foods *(facing page)*.

Asparagus Etiquette

Perhaps because it is so temptingly a "finger food," people concerned with table manners have gone back and forth regarding the proper way to eat unadorned asparagus spears. Sometime in the early 1900s, everyone got the go-ahead to use the fingers, even in elegant surroundings. Then there's the tale of a foreign visitor to Paris who ate his asparagus by biting off the tips and tossing the offending ends over his shoulder. Amazed though his hosts were, they did not want to make him feel uncomfortable. So they did the same. (Try this either at home or at a restaurant and you'll be in big trouble . . .)

serve cold steamed asparagus with an oil and vinegar dressing or bake it in egg pies called quiches or sauté it lightly in oil with other vegetables such as red peppers.

Hungarians like something called sparga pudding—asparagus cooked with beef broth and sour cream. Asparagus baked with eggs, cheese, and green onions is a favorite dish of Armenia in Central Asia, along with asparagus fritters. Greeks serve warm or cold asparagus with a dressing made from lemon, olive oil, capers (the buds of a Mediterranean shrub), and finely chopped herbs such as basil, marjoram, and rosemary. An Italian specialty features asparagus spears and fontina cheese wrapped in paper-thin slices of a fine ham called prosciutto and then baked.

It's a Fact!

The first known cookbook in the world is entitled *Of Culinary Matters*. (*Culinary* means "in regard to cooking" and comes from *culina*, the Latin word for kitchen.) A Roman named Apicius wrote the book in A.D. 14. He suggested combining asparagus with herbs, wine, and eggs to make a kind of unsweetened, oven-baked pudding.

True asparagus fans have more than one U.S. festival to attend. In California the Stockton Asparagus Festival held each April offers asparagus with everything—asparagus burritos, asparagus nachos with asparagus salsa, deep-fried asparagus, asparagus pasta, and asparagus bisque (a thick, creamy soup). In Shelby, Michigan, the National Asparagus Festival in June also pushes the boundaries of asparagus cuisine. Festival-goers can lap up asparagus guacamole with pieces of asparagus bread and finish up with an asparagus-rhubarb pie for dessert.

Volunteers at the Stockton Asparagus Festival in Stockton, California, serve up all kinds of tasty asparagus treats.

Asparagus Fritters (2–3 servings)

1 cup asparagus pieces
1 cup flour
1 teaspoon baking powder
½ teaspoon salt
1 egg
½ cup milk
1 teaspoon vegetable oil, plus more for frying

If you're using fresh asparagus, then lightly grasp a spear just below the tip and at the base of the stalk. Bend the spear until it snaps. Discard the tough stalk end; you'll only use the portion with the bud. Cut the asparagus into 1½-inch pieces and toss them into a pot of boiling water. Cook about one minute and then remove the asparagus with a slotted spoon. While the asparagus is cooling, combine the other ingredients in a large mixing bowl. Mix the asparagus into the flour batter.

You may want to ask an adult to help with the frying. Pour about half an inch of oil into a heavy pan and heat it at a medium setting. Drop spoonfuls of the asparagus mixture into the pan, being careful not to splash yourself with the oil. Cook the fritters until golden brown, turning them over once. Remove the fritters with a slotted spoon and let drain briefly on paper towels. Eat at once.

Rhubarb

[*Rheum rhaponticum*]

How is it that rhubarb, a plant whose leaves are deadly poisonous, nonetheless grows happily in home gardens and turns up in favorite summer desserts? The story is a long one, as people have been using rhubarb for one thing or another for ages.

Rhubarb started out its long career as a medicinal food, far from a tasty dessert. An ancient plant with origins unknown, it may have come from Asia, possibly northern China, where it turns up as an important entry in an **herbal** called *Pen-king* in about 2700 B.C.

At first the rhubarb's root was the source of its appeal, and Chinese traditional medicine recommended it for constipation and digestive troubles. Centuries later the Romans used rhubarb for the same complaints, and they also gave the plant its name. They combined the words *Rha*—the name of the river along which the plant grew (these days called the Volga in western Russia)—and *barbarum*, or foreign, because territory beyond that river didn't belong to the Romans.

Harvested rhubarb waits in a field to be picked up and shipped to a supermarket.

Even the tongue and teeth have a scrubbed feeling after a dish of early rhubarb.

—Della T. Lutes

ROVIRA SOLSONA, S.L.
TELFS.-964-511827 y 673273
PROD. EN ESPAÑA
R.E.12.232120

Marco Polo, a merchant from Venice, Italy, journeyed in Asia for several years beginning in A.D. 1275. After seeing rhubarb root in China, Polo wrote that it was exported "far and wide." Also in the 1200s, the Arab chronicler Ibn el-Beithar noted that rhubarb was "common in Syria and Persia" (present-day Iran). The plant eventually arrived in the hands of European herbalists during the Middle Ages, again in root form, to be used as medicine.

By this time, European monks—who were the healers in most communities—were growing rhubarb in monastery gardens. The root was considered so valuable that European governments tried to get sole rights to its sale and distribution. In 1772 Russia finally succeeded in establishing a deal with a company in the town of Lanzhou on the Chinese frontier. Caravans carried the healing root to St. Petersburg, which was then the Russian capital, and from there merchants took it throughout Europe and sold it as Royal Rhubarb.

Medieval druggists knew of rhubarb's healing properties.

In about 1777, a British apothecary (druggist) known to us only as Hayward began growing rhubarb from seeds he had received from Russia 15 years earlier. He raised several different varieties, including the one we eat.

A British drawing of rhubarb from around 1798

Go Ahead, Be A Blond

If you have blondish or even mousy hair, a rhubarb rinse could be your ticket into the true land of blond. First check with your parents to make sure they don't mind. Then find a source of rhubarb root, perhaps at your local natural-foods store. Put three tablespoons of grated root in a pot with two cups of water and boil gently for about 15 minutes. Set aside overnight. The next day, strain through a sieve. Pour the liquid through your hair as a rinse after shampooing. (Of course you might want to test a few strands first, to see if you like the look.)

Valued exclusively for its root, rhubarb was not prized for its stalks until the early 1800s—perhaps because the first people to eat other parts of the rhubarb plant ate its leaves, which are highly toxic. It is likely that people experimenting with new ways to use the plant got sick, and some died. It's no wonder that rhubarb, other than in root form, was not wildly popular.

Soon rhubarb found its way to North America. An unknown gardener from Maine is said to have obtained rhubarb stock from Europe in about 1800. He grew it and in turn introduced it into Massachusetts, where by 1822 growers were selling rhubarb in farmers' markets.

Many people grow rhubarb in their yards because, with minimal care, the plant provides stems all summer long.

A commercial rhubarb plant's stalks are ready for harvest.

Growing the Pink Stalk

Rhubarb is a cool-weather, perennial plant. Like asparagus, it requires a winter dormancy period to stimulate the next season's growth. Commercial growers plant rhubarb in both Canada and the United States, but on a very small scale compared to other vegetable crops. The key U.S. producers are Washington, Oregon, and Michigan, but these states' commercial operations total 300 acres or less per state. Together they raise 25 to 30 million tons of rhubarb annually. Poland, Russia, and the United Kingdom also grow rhubarb commercially.

Rhubarb is usually grown from crowns placed in the ground about 24 to 36 inches apart. Growers plant each crown in a large hole filled with organic matter and give each plant plenty of water. Growers must wait until the root system is well developed to ensure good productivity from each plant. The tall, red stalks are ready for harvest starting in the second or third year. The rhubarb can be cut at soil level or even pulled out of the ground. New stalks keep sprouting, so several harvests are available each season. After a few years of yield, each crown should be dug up and divided to ensure strong stalks on future plants. With care a rhubarb crown can produce crop for up to 20 years.

Commercial greenhouse rhubarb growers leave the plants in the field for two or three years without harvesting to produce root clumps called mother plants. Farmers bring the mother plants into greenhouses. The clumps receive a spray of growth hormones, and the stalks begin to shoot up. The growers harvest the mother plants regularly over three months and then throw them away.

Rubarb stalks destined for the fresh market are trimmed and packed into 15- or 20-pound cartons. Packers don't wash the rhubarb, because the water could cause the stalks to spoil rapidly. The big boxes go to supermarkets, where grocers set out bundles of rhubarb for shoppers. Crop for processing is washed, chopped, quick-frozen, and packed into plastic bags. Growers send this rhubarb to makers of pies and condiments.

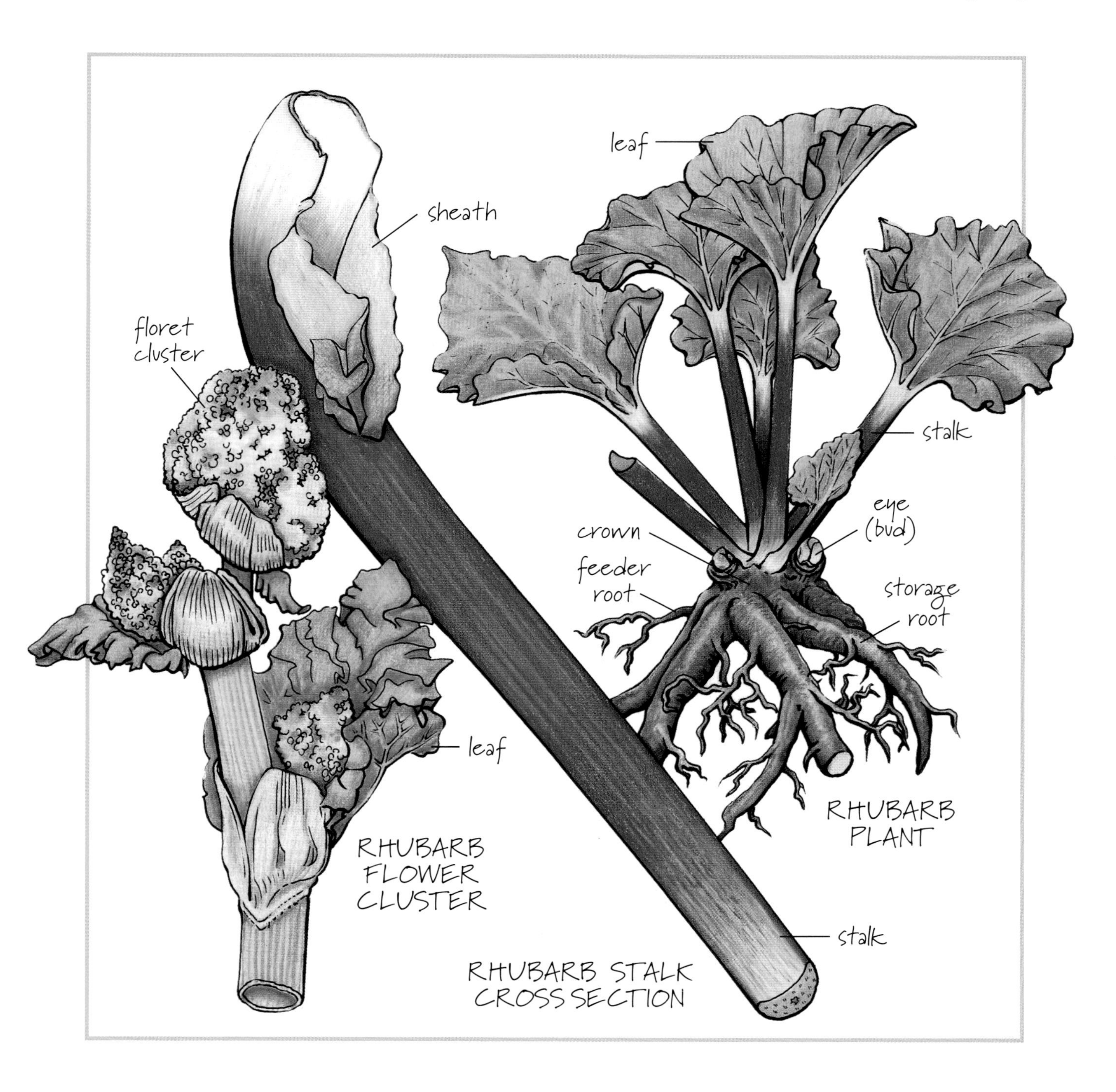

sheath
floret
cluster
leaf
leaf
RHUBARB
FLOWER
CLUSTER
stalk
RHUBARB STALK
CROSS SECTION
leaf
stalk
crown
eye
(bud)
feeder
root
storage
root
RHUBARB
PLANT

Juicy strawberry-rhubarb pie is a well-loved summertime treat.

Relishing Rhubarb

If you've never eaten fresh strawberry-rhubarb pie, you have something delicious to anticipate. Whoever first paired rhubarb with strawberries should be given an award. The combination is pleasing in a variety of dishes, including gelatin desserts, puddings, egg-filled soufflés, and jams, as well as pie. Such dishes are usually considered North American or British treats, although Scandinavians also frequently use these mixtures.

In Britain, particularly in Cornwall, cooks top grilled mackerel, a flavorful Atlantic fish, with sugared rhubarb in white sauce (a type of gravy made from milk, flour, and butter). In nineteenth-century Britain, rhubarb fritters were a tasty treat. Stalks of young rhubarb were dipped in batter, deep-fried for about five minutes, and then sprinkled with sugar.

Famous for homey desserts, the British find many uses for rhubarb.

The Swiss-born chef Albert Stockli was enthusiastic about rhubarb and felt it could stand alone. In the 1960s, when he was devising recipes for New York's famous restaurant, the Four Seasons, he decided to

The Four Seasons restaurant, birthplace of Rhubarb Stockli

Rhubarb Crunch (4–6 servings)

- 1½ cup flour, sifted
- 1½ cup oatmeal
- 1 cup brown sugar
- ½ cup melted butter
- 1 tablespoon cinnamon
- 4 cups diced rhubarb
- 1 cup white sugar
- 2 tablespoons cornstarch
- 1 cup water
- 1 teaspoon vanilla

Mix the first five ingredients together until they have a crumbly texture. Press half of the mixture into a 9x9-inch pan to form a bottom crust. Place the rhubarb in the pan.

Put the last four ingredients into a saucepan and stir them together. Cook, stirring constantly, for several minutes, until the liquid is clear. Pour the sauce over the rhubarb and then top with the remaining crumbs. Bake at 350° for one hour.

create a strawberry-free rhubarb dessert. The result was Rhubarb Stockli, a rich, sweet pudding. The secret extra ingredients were vermouth and Madeira, two flavorful wine drinks. Rhubarb itself has frequently been made into a homemade wine. An herbal after-dinner drink in Italy features rhubarb as a key ingredient.

Icelanders have long used rhubarb on its own to make soup and pudding. Each recipe calls for the same four ingredients—rhubarb, water, sugar, and potato flour. The pudding uses less water and more flour. In Poland people combine rhubarb with potatoes, cooking them in a rich onion and mushroom broth—a tasty side dish to meat.

Rhubarb to the Rescue

Rhubarb leaves contain oxalic acid, a substance that occurs in tiny amounts in our bodies. (Even so, oxalic acid can sicken or kill you if you eat enough of it, so don't ever eat rhubarb leaves.) Oxalic acid is found in very small quantities in other vegetables, including spinach, cabbage, beet greens, peas, and potatoes. The oxalic acid in rhubarb leaves may combine with another compound, something called anthraquinone glycosides, to cause the high toxicity not found in the other plants.

Rhubarb also works well as a pot scrubber for tarnished metals, using either the leaves or the stalks.

Boiled up in a quart of water and garnished with a few drops of liquid soap, however, the leaves make a fine aphid spray. (Aphids are tiny insects that can cause great damage to crops and houseplants.) That's oxalic acid working for good.

And oxalic acid may help save the earth's ozone layer, which protects the earth from the sun's potent ultraviolet rays. Human-made substances called chlorofluorocarbons (CFCs), once used as coolants in appliances, are causing holes in the ozone layer. But passing the CFCs through a heated bed of an oxalic acid compound breaks them down into four harmless components—sodium chloride or table salt; sodium fluoride, an ingredient in toothpaste; carbon; and carbon dioxide. New laws prohibit CFCs in appliances, but there are plenty of old fridges and air conditioners lying around in dumps, waiting for their CFCs to be destroyed.

Rhubarb could help clean up the CFCs in this dump.

Glossary

blanch: To block sunlight from reaching a plant, thereby preventing photosynthesis from taking place and keeping the plant pale in color.

bulb: An underground growth consisting of a short stem and one or more buds wrapped tightly by specialized, fleshy leaves.

calorie: A unit of measurement expressing the amount of heat produced by a food when it burns. Scientists use this information to determine how much energy a food provides when it is fully digested and used by the body.

crown: A first-year growth of roots and underground stems that can be planted instead of seeds to grow some perennial crops.

domestication: Taming animals or adapting plants so they can safely live with or be eaten by humans.

dormant: Not growing, but capable of growing later.

herbal: An early book that describes plants and their many uses, frequently for healing.

organic matter: Dead plants and animals in various stages of rotting.

photosynthesis: The chemical process by which green plants make energy-producing carbohydrates. The process involves the reaction of sunlight to carbon dioxide, water, and nutrients within plant tissues.

tropics: A zone of the world, shaped like a belt around the globe's midsection, having a hot, humid climate and often heavy and diverse plant growth.

Further Reading

Hill, Lee Sullivan. *Farms Feed the World.* Minneapolis: Carolrhoda Books, 1997.

Nottridge, Rhoda. *Vitamins.* Minneapolis: Carolrhoda Books, 1993.

Root, Waverly. *Food.* New York: Simon & Schuster, 1980.

Trager, James. *The Food Chronology.* New York: Henry Holt and Company, 1995.

Vegetarian Cooking around the World. Minneapolis: Lerner Publications, 1992.

Wake, Susan. *Vegetables.* Minneapolis: Carolrhoda Books, 1990.

A woman examines a bundle of scallions at an open-air market in China.

Index

activities, 17, 48, 57
allium plants, 8, 10, 28, 32, 40
Asia, 10, 14, 27, 34, 50, 76, 78
asparagus, 60–75; growing, 64, 66–69; harvesting, 66–67; health information, 71; origins, 60, 62–63; recipe, 75; uses for, 70, 72–74; white, 68

Babylonians, 14
Bermuda onion, 17–18
Black Death, 31
blanching, 47, 52, 56, 68
British Isles, 43–44, 50, 62

Caribbean, 18, 25, 34
celeriac, 53
celery, 50–59; growing, 53–56; harvesting, 56; health information, 58; origins, 50–52; uses for, 54, 58–59
Celts, 43–44
Chicago, 16–17
Chile, 19
China, 14, 28, 53–54, 59, 68–69, 76, 78
chlorofluorocarbons, 84
Cook, Captain James, 16
Cortés, Hernán, 16

de Bruin, Cornelius, 54

Egypt, 13, 42
England, 16, 25, 43

farming, 21–24, 37–38, 47, 53–54
France, 31, 33–34, 44, 52–53, 59, 62–63

garlic, 28–39; harvesting, 37–38; health information, 31, 35; origins, 28–32; planting, 37; recipe, 39; uses for, 33–35
Gilroy Garlic Festival, 33, 35
Great Britain, 30, 82
Greeks, 30, 33, 42, 52, 63

harvesting, 23–24, 37–38, 47, 56, 66–67, 80
health information, 13, 27, 31, 35, 42, 58, 71, 76, 78

Italy, 31–33, 52

Kalamazoo, Michigan, 54

leeks, 40–49; harvesting, 47; health information, 42; origins, 40–44; planting, 47–48; recipe, 49; uses for, 44–45

Maui onions, 17–19
Mediterranean, 10, 14, 26, 28, 40, 50, 60, 62
Mexico, 16
Middle Ages, 15, 31, 52, 78
Middle East, 10, 12–13, 26, 28, 34, 40, 78

National Asparagus Festival, 74
Navajo, 17
North America, 10, 16, 18–19, 32–33, 47, 53, 56, 79

onions, 10–27; harvesting, 21–23; health information, 13, 27; origins, 10, 12–17; planting, 21–22; recipe, 26; storing, 23–24; uses for, 24–27

oxalic acid, 84

pearl onions, 20
Peru, 68–69
photosynthesis, 4–6, 47, 68
planting, 21, 37, 47–48, 64, 66, 80
plants: bulb and stem vegetables, 7–8; domestication, 7; origins, 6–7;photosynthesis, 4
Pompeii, 14
pyramid of Cheops, 13

recipes, 26, 39, 49, 59, 75, 83
rhubarb, 76–84; growing, 80–81; health information, 76, 78; origins, 76–79; recipe, 83; uses for, 79, 82–84
Roman Empire, 14–15, 30–31, 42, 50, 52, 62–63, 76

Saint David, 44
Saint Patrick, 42
scallions, 20
scurvy, 13, 16
shallots, 7–8, 20
Spain, 16–17, 31–34, 68
Stockton Asparagus Festival, 74
Sumeria, 12

taxonomy, 12

United States, 16, 18–19, 24, 32, 37, 53, 68, 80

Vidalia onions, 17, 19

Wales, 43–44
Walla Walla onions, 17–18

About the Author

Meredith Sayles Hughes has been writing about food since the mid-1970s, when she and her husband, Tom Hughes, founded The Potato Museum in Brussels, Belgium. She has worked on two major exhibitions about food, one for the Smithsonian and one for the National Museum of Science and Technology in Ottawa, Ontario. Author of several articles on food history, Meredith has collaborated with Tom on a range of programs, lectures, workshops, and teacher-training sessions, as well as on *The Great Potato Book*. The Hugheses do exhibits and programs as The FOOD Museum in Albuquerque, New Mexico, where they live with their son, Gulliver.

Acknowledgments

For photographs and artwork: Steve Brosnahan, p. 5; TN State Museum, detail of a painting by Carlyle Urello, p. 7; DDB Stock Photo/Suzanne Murphy-Larronde, p. 11; Erich Lessing/Art Resource, NY, p. 13; © Betty Crowell, p. 14, 21 (bottom), 80; Independent Picture Service, p. 15, 78 (right); © Grant H. Kessler, p. 16; Gary Holscher/© Tony Stone Images, p. 18; © Bob Bass Photo, p. 19; © Craig D. Wood, p. 20 (both), 29, 39 (inset), 57 (bottom), 59, 71; Gavin Hellier/© Tony Stone Images, p. 21 (top); © Betty Derig/Photo Researchers, 23 (left); Inga Spence/TOM STACK & ASSOCIATES, p. 23 (right); Meredith Sayles Hughes, p. 25; © September 8th Stock, Walt/Louiseann Pietrowicz, p. 26, 27, 39, 44 (left), 45 (both), 51, 58, 72, 75, 83; North Wind Picture Archives, p. 30, 44 (right); Giraudon/Art Resource, NY, p. 31; Hollywood Book and Poster, p. 32; © Bruce Berg/Visuals Unlimited, p. 33 (top); © Dr. Roma Hoff, p. 33 (bottom); © Robert Fried, p. 34, 38, 86; © Alan & Linda Detrick/Photo Researchers, p. 35 (left); © Mark E. Gibson/Visuals Unlimited, p. 35 (right); Chad Slattery/© Tony Stone Images, p. 37 (top); © Michael P. Gadomski/Photo Researchers, p. 37 (bottom), 64 (inset); Holt Studios/Nigel Cattlin, p. 41, 56 (right); Archive Photos, p. 42, 62; Britain on View, p. 43; Holt Studios/Willem Harinck, p. 47 (top); Holt Studios/Richard Anthony, p. 47 (bottom), 77; © Stephen Daniels, p. 48; Robert L. Wolfe, p. 49; Mary Evans Picture Library, p. 52; © D. Cavagnaro/Visuals Unlimited, p. 53; © Frederick Myers/Tony Stone Images, p. 56 (left); DDB Stock Photo/John Vivian, p. 57 (top); DDB Stock Photo/Inga Spence, p. 61; © John D. Cunningham/Visuals Unlimited, p. 63; © SuperStock, Inc., p. 64, 70; Michigan Asparagus Advisory Board, p. 66, 67 (both), 73; © Warren Stone/Visuals Unlimited, p. 68; © Sylvan H. Wittwer/Visuals Unlimited, p. 69; Martin Brown/Stockton Convention and Visitors Bureau, p. 74; Image Select/Art Resource, NY, p. 78 (left); Holt Studios/Gordon Roberts, p. 79; Jim Simondet/Independent Picture Service, p. 82 (top); The Four Seasons, p. 82 (bottom); Jerry Boucher, p. 84. Sidebar and back cover artwork by John Erste. All other artwork by Laura Westlund. Cover photo by Steve Foley and Réna Dehler.

For quoted material: p. 4, M.F.K. Fisher, *The Art of Eating* (New York: Macmillan Reference, 1990); p. 40, Geoffrey Chaucer, "The Summoner's Tale," *The Canterbury Tales* (1343?–1400); p.50, Ogden Nash, as quoted by Joan and John Digby, eds., *Food for Thought* (New York: William Morrow & Company, 1987); page 60, as quoted by Michael Cader, ed., *Eat These Words,* (New York: Harper Collins, 1991); page 76, Della T. Lutes, *The Country Kitchen* (Boston: Little, Brown, and Co.: 1936).

For recipes (some slightly adapted for kids): p. 26, Jan Roberts-Dominguez, *The Onion Book* (New York: Doubleday, 1996); p. 39, Meredith Sayles Hughes; p. 49, Lynne Marie Waldee, *Cooking the French Way* (Minneapolis: Lerner Publications Company, 1982); p. 75, Sonia Uvezian, *The Cuisine of Armenia* (New York: Hippocrene Books, 1996); p. 83, Washington Rhubarb Growers Association.